SpringerBriefs in Physics

SpringerBriefs in Physics are a series of slim high-quality publications encompassing the entire spectrum of physics. Manuscripts for SpringerBriefs in Physics will be evaluated by Springer and by members of the Editorial Board. Proposals and other communication should be sent to your Publishing Editors at Springer.

Featuring compact volumes of 50 to 125 pages (approximately 20,000-45,000 words), Briefs are shorter than a conventional book but longer than a journal article. Thus, Briefs serve as timely, concise tools for students, researchers, and professionals.

Typical texts for publication might include:

- A snapshot review of the current state of a hot or emerging field
- A concise introduction to core concepts that students must understand in order to make independent contributions
- An extended research report giving more details and discussion than is possible in a conventional journal article
- A manual describing underlying principles and best practices for an experimental technique
- An essay exploring new ideas within physics, related philosophical issues, or broader topics such as science and society

Briefs allow authors to present their ideas and readers to absorb them with minimal time investment.

Briefs will be published as part of Springer's eBook collection, with millions of users worldwide. In addition, they will be available, just like other books, for individual print and electronic purchase.

Briefs are characterized by fast, global electronic dissemination, straightforward publishing agreements, easy-to-use manuscript preparation and formatting guidelines, and expedited production schedules. We aim for publication 8-12 weeks after acceptance.

More information about this series at http://www.springer.com/series/8902

Luís Manuel Couto Oliveira •
Valery Victorovich Tuchin

The Optical Clearing Method

A New Tool for Clinical Practice and
Biomedical Engineering

 Springer

Luís Manuel Couto Oliveira
Physics Department and Center
for Innovation in Engineering
and Industrial Technology
Polytechnic Institute
of Porto – School of Engineering
Porto, Portugal

Valery Victorovich Tuchin
Department of Optics and Biophotonics
Saratov State University
Saratov, Russia

Institute of Precision Mechanics and Control
of the RAS
Saratov, Russia

Bach Institute of Biochemistry
Research Center of Biotechnology of the RAS
Moscow, Russia

Tomsk State University, Tomsk & ITMO
University
St. Petersburg, Russia

ISSN 2191-5423   ISSN 2191-5431   (electronic)
SpringerBriefs in Physics
ISBN 978-3-030-33054-5   ISBN 978-3-030-33055-2   (eBook)
https://doi.org/10.1007/978-3-030-33055-2

This Springer imprint is published by the registered company Springer Nature Switzerland AG
The registered company address is: Gewerbestrasse 11, 6330 Cham, Switzerland

Preface

The optical immersion clearing method allows one to increase tissue transparency through the reduction of light scattering. The method has considerable potential that has captured researcher's interest in the last two decades and produced an increasing number of publications in the past few years. In recent years, this technique has been applied in various in vitro, ex vivo, and in vivo biological materials, allowing, among others, the evaluation of the magnitude of the transparency effects, evaluation of mobile water in tissues, estimation of the diffusion properties of agents in biological tissues and liquids, discrimination between normal and pathological tissues, and characterization of the mechanisms involved in the optical clearing process. Considering that more than 10 years have passed since the first monograph has been published about the optical clearing method, we decided to write the present monograph to review and discuss some novel aspects of this fast growing research field.

This book covers the most recent discoveries in the field of optical clearing of biological tissues, the data that can be acquired from experimental studies, and its application in many fields of clinical practice or biomedical engineering. Chapter 1 describes the problem of tissue scattering and the limitations to the use of light-based methods in clinical practice. Chapter 2 describes various methods to increase tissue transparency and presents an historical description of the optical immersion clearing method. Some typical optical clearing agents and their optical properties are presented in Chap. 3, while the mechanisms of optical clearing are described in Chap. 4. The measurements that can be performed from ex vivo or in vivo tissues during optical clearing treatments are described in Chap. 5, and the data that can be retrieved from those measurements is described in Chap. 6.

Another very important field is tissue imaging. The most recent developments in tissue clearing that are directly connected to tissue imaging are presented in Chap. 7. The particular application of tissue and organ preservation is intimately connected with the optical clearing method, since most optical clearing agents are also used as cryoprotective agents. Data that can be acquired during treatments, and are useful for other areas of biomedical engineering, such as tissue preservation, are presented in Chap. 8. Chapter 9 discusses the future of optical clearing in clinical practice, tissue and organ preservation, food industry, and other fields of biomedical engineering.

Porto, Portugal   Luís Manuel Couto Oliveira
Saratov, Russia   Valery Victorovich Tuchin

Acknowledgments

Luís Oliveira is thankful for support from the Portuguese Foundation for Science and Technology through grant no. UID-EQU-04730-2019.

Valery V. Tuchin is thankful for support from grants of RFBR 17-02-00358, KOMFI 17-00-00275 (17-00-00272), 18-52-16025 NTSNIL_a, 18-29-02060 MK; the RF Presidential grant 14.Z57.16.7898-NSh; the RF Governmental grants 08-08, 14. Z50.31.0004, 14. Z50.31.0044, and 14.W03.31.0023; the RF Ministry of Science and Higher Education 17.1223.2017/AP; and by the Basic Research Program of the Presidium of the RAS N32.

Both authors are thankful to the colleagues of the Portuguese Oncology Institute of Porto, Portugal, for collecting and preparing the biological tissues used in some of the research presented in this monograph.

The authors would like to show their appreciation to the colleagues of the Department of Optics and Biophotonics of Saratov State University for collaboration and especially to the PhD student Ekaterina Lazareva for measuring the refractive index of some optical clearing agents. The authors are thankful to Dr. Viacheslav Artyushenko and Dr. Olga Bibikova from art photonics GmbH in Germany for providing the diffuse reflectance sensor that was used to acquire measurements that are presented in Chap 5.

We express our gratitude to our families for their indispensable support, understanding, and patience during the writing of this book.

Abbreviations

C$_e$3D	Clearing-enhanced 3D
CEES	Chloroethyl ethyl sulfide
CLSM	Confocal light-sheet microscope
CM	Confocal microscopy
CPA	Cryoprotective agent
CPE	Chemical penetration enhancers
CSF	Cerebral spinal fluid
CT	Computed tomography
CUBIC	Clear unobstructed brain imaging cocktails
DMSO	Dimethyl sulfoxide
ECi	Ethyl cinnamate
e-Cig	Electronic cigarette
EG	Ethyleneglycol
ETC	Electrophoretic tissue clearing
FPT	Fructose + PEG-400 + thiazone solution
Hct	Hematocrit
HF	Hydrofluoric acid
HO	Hyperoxia exposure
IAD	Inverse adding doubling
IMC	Inverse Monte Carlo
ISF	Interstitial fluid
LSCI	Laser speckle contrast imaging
LSFEM	Light-sheet fluorescence expansion microscopy
MPT	Multiphoton tomography
MRI	Magnetic resonance imaging
NA	Numerical aperture
NaCl	Sodium chloride
NADPH	Nicotinamide adenine dinucleotide phosphate
NIR	Near infrared
OC	Optical clearing

OCA	Optical clearing agent
OCT	Optical coherence tomography
ODD	Optical detection depth
P	Permeability rate
PBS	Phosphate buffered saline
PEG	Polyethylene glycol
PEGASOS	Polyethylene glycol associated solvent system
PET	Positron emission tomography
PG	Propylene glycol
PMT	Photomultiplier tube
RI	Refractive index
RIMS	Refractive index matching solution
RPE	Retinal pigment epithelium
RSDL	Reactive skin decontamination lotion
SC	Stratum corneum
SD	Standard deviation
SDS	Sodium dodecyl sulfate
SeeDB	See deep brain
SHG	Second harmonic generation
SLED	Super-luminescent light emitting diode
SLS	Sodium lauryl sulfate
SPIM	Selective plane illumination microscopy
TDE	2,2′-thiodiethanol
TELS	Tissue engineered liver slices
UV	Ultraviolet
VF	Volume fraction

Contents

Chapter 1
Tissue Optics

1.1 Optical Properties of Tissues and Light Propagation

When light travels inside a medium, it suffers attenuation, which occurs through two mechanisms—absorption and scattering. The amount of the incident light that is absorbed or scattered inside a material is quantified by the characteristic optical properties of the material—the absorption coefficient (μ_a) and the scattering coefficient (μ_s) [1]. Absorption and scattering of light occur when the photons in the incident beam interact with particles inside the material. To understand both these interactions and define μ_a and μ_s, we consider a collimated beam, in which path a particle was placed.

In Fig. 1.1, σ_a and σ_s represent the effective absorption and scattering cross sections of the particle. These sections are two-dimensional areas that are perpendicular to the beam direction.

From Fig. 1.1a we see that some of the photons in the beam are directed to σ_a (blue rays) and others pass around it (black). If there are many absorbing particles in the beam path with a density ρ_a, then the absorption coefficient is defined as [2]:

$$\mu_a = \rho_a \times \sigma_a. \tag{1.1}$$

Since σ_a is an area that is usually represented in cm^2 and ρ_a is represented in cm^{-3}, then μ_a is presented in cm^{-1}. This means that μ_a is a measure of the number of photons that are absorbed in the medium per unit length of the beam path.

Similarly, considering Fig. 1.1b, we see that the photons that interact with σ_s (green rays) are scattered to other directions. Again, considering a density of scattering particles in the beam path (ρ_s), the scattering coefficient is defined as [2]:

$$\mu_s = \rho_s \times \sigma_s. \tag{1.2}$$

© The Author(s), under exclusive license to Springer Nature Switzerland AG 2019
L. M. C. Oliveira, V. V. Tuchin, *The Optical Clearing Method*,
SpringerBriefs in Physics, https://doi.org/10.1007/978-3-030-33055-2_1

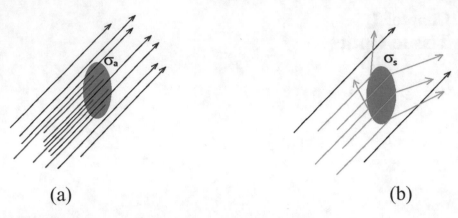

(a)   (b)

Fig. 1.1 Photon absorption (**a**) and photon scattering (**b**)

In analogy to μ_a, μ_s is also a measure of the number of photons that are scattered in the medium per unit length of the beam path. The most common unit to represent these coefficients is cm^{-1}. Knowing μ_a and μ_s, the total attenuation of the medium (μ_t) is simply calculated as [2]:

$$\mu_t = \mu_a + \mu_s. \tag{1.3}$$

When light scattering occurs, any direction can be taken by the scattered photon, depending on the photon wavelength used and the size of the scatterers in the medium. The probability of a photon traveling in the direction $\vec{s}$ to be scattered into a direction $\vec{s}'$ is calculated by the phase function—$p\left(\vec{s}, \vec{s}'\right)$ [1]. When scattering is symmetric around the initial direction of the beam, then the phase function only depends on the scattering angle θ, between directions $\vec{s}$ and $\vec{s}'$. This means that $p\left(\vec{s}, \vec{s}'\right) = p(\theta)$, or in other words, it only depends on the angle between the photon directions before and after the scattering event. In practice, the phase function is well approximated by the Henyey-Greenstein function [1, 3]:

$$p(\theta) = \frac{1}{4\pi} \frac{1 - g^2}{(1 + g^2 - 2g \cos\theta)^{3/2}}, \tag{1.4}$$

where g represents the scattering anisotropy parameter (mean cosine of the scattering angle θ) [1, 3]:

$$g \equiv <\cos\theta> = \int_0^\pi p(\theta) \cdot \cos\theta \cdot 2\pi \cdot \sin\theta\, d\theta. \tag{1.5}$$

Depending on the scattering properties of the material, g values range from -1 to 1. Isotropic scattering corresponds to $g = 0$, and total forward scattering is characterized by $g = 1$ and total backward scattering by $g = -1$ [1, 3].

Another optical property is the reduced scattering coefficient, μ_s'. This property provides a certain measure of scattered photons directionality, since it is related both with μ_s and g [1, 3, 4]:

$$\mu_s' = \mu_s \times (1 - g). \tag{1.6}$$

Another interesting property that can also be calculated from the previous is the light penetration depth (δ). It measures how deep light can penetrate into a medium before reaching 37% of its original intensity. A simple way to calculate δ from μ_a and μ_s' can be made in diffusion approximation [5, 6]:

$$\delta = \frac{1}{\sqrt{3\mu_a(\mu_a + \mu_s')}}. \tag{1.7}$$

The importance of δ is high, for instance, for the correct calculation of the light irradiation dose in photothermal and photodynamic therapy of various diseases [1].

The above optical properties can be used to calculate others [2], but the previous ones are the most fundamental and most important in tissue optics.

Finally, a last optical property that is also very important is the refractive index (RI). The RI is a parameter that quantifies the light speed in a medium and how light direction changes between media [2]. Considering c as the light speed in vacuum ($c = 2.998 \times 10^8$ m/s) and v the light speed in a biological tissue, tissue RI (n_{tissue}) is calculated as:

$$n_{\text{tissue}} = \frac{c}{v}. \tag{1.8}$$

Considering biological tissues, all the above presented optical properties depend on the wavelength. This means that to perform a diagnosis or treatment, the wavelength of the light to be used must be carefully selected. Tissues are heterogeneous materials and have various components. Consequently, the global optical properties of a tissue are a combination of the individual properties of tissue's components. Some of the most common tissue components are water, blood, proteins, and lipids. These biological components have some important absorption bands that condition the light selection to some particular wavelength bands. The μ_a spectra of water, lipids, hemoglobin, and other chromophores for the range between 200 and 1200 nm are presented in Fig. 1.2 [7–11].

The spectra for μ_s must also be considered to select an appropriate wavelength of the light to use in clinical procedures. In general, biological tissues have μ_s values much higher than the μ_a values. The wavelength dependence for μ_s (and for μ_s', as well) decreases with wavelength in the visible-NIR range and represents a

Fig. 1.2 Absorption spectra
(μ_a) from some tissue
components. (Adapted from
Ref. [7])

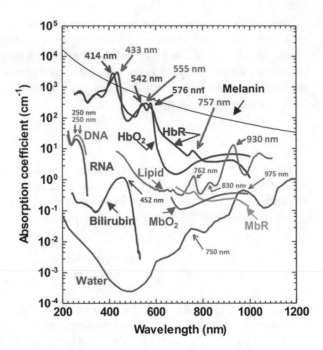

combination between Rayleigh and Mie scattering terms at the reference wavelength. Such behavior is described by Eq. (1.9) [4]:

$$\mu_s(\lambda) = a' \left(f_{\text{Ray}} \left(\frac{\lambda}{500 \ (\text{nm})} \right)^{-4} + (1 - f_{\text{Ray}}) \left(\frac{\lambda}{500 \ (\text{nm})} \right)^{-b_{\text{Mie}}} \right). \qquad (1.9)$$

In Eq. (1.9), the scaling factor a' represents the μ_s at 500 nm. This equation has
been successfully used to fit the wavelength dependence for data of many biological
soft tissues (see Table 1 of Ref. [4]).

To show that μ_s is much higher than μ_a, we have grouped a few data values for
some particular biological tissues in Table 1.1.

The data presented in Table 1.1 shows that light scattering is much higher than
absorption in biological tissues and blood. This is a characteristic of natural biological materials. Considering the visible-NIR range and the decreasing behavior with
wavelength for μ_s, as described by Eq. (1.9), such difference increases with decreasing wavelength. The graphs in Fig. 1.3 show such behavior for human colorectal
tissues [19, 20].

The data in graphs of Fig. 1.3 were estimated using the inverse adding-doubling
(IAD) method [21]. By comparing between graphs of Fig. 1.3a, b, we see that all
tissues studied have μ_s values much higher than the μ_a values.

The curves in graphs of Fig. 1.3b, c were obtained by fitting the discrete data and
are described by Eqs. (1.10)–(1.15) [19, 20].

Table 1.1 Absorption and scattering coefficients of some biological tissues

Tissue	λ (nm)	μ_a (cm^{-1})	μ_s (cm^{-1})	Observations	References
Skin (human)	337	32	165	Epidermis 100 μm thick samples (IMC simulations)	[12, 13]
	577	10.7	120		
	633	4.3	107		
	337	23	227	Dermis 200 μm thick samples (IMC simulations)	
	577	3.0	205		
	633	2.7	187		
Muscle (Wistar Han rat)	400	7.2	119	Skeletal muscle 500 μm thick samples (Average between IMC and IAD simulations)	[14]
	500	4.1	106		
	600	2.1	100		
	700	1.1	96.5		
	800	1.1	94		
	900	1.1	92		
	1000	1.2	91		
Liver (human)	850	1	204	200 μm thick samples (IMC simulations)	[15]
	980	0.8	182		
	1064	0.5	169		
Brain (human)	450	1.4	420	White matter 80–150 μm thick samples (IMC simulations)	[16]
	510	1.0	426		
	630	0.8	409		
	670	0.7	401		
	850	1.0	342		
	1064	1.0	296		
	450	0.7	117	Gray matter 100–200 μm thick samples (IMC simulations)	
	510	0.4	106		
	630	0.2	90		
	670	0.2	84		
	1064	0.5	57		
	400	4.7	276.7	Cerebellum 100–200 μm thick samples (IMC simulations)	
	500	1.4	277.5		
	600	0.8	272.1		
	700	0.6	266.8		
	800	0.6	250.3		
	900	0.7	229.6		
	1000	0.8	215.4		
	1100	0.7	202.1		
Blood (human)	665	1.3	1246	Whole blood HbO$_2$ (Hct = 0.41) (IMC simulations)	[17, 18]
	685	2.65	1413		
	960	2.84	505		

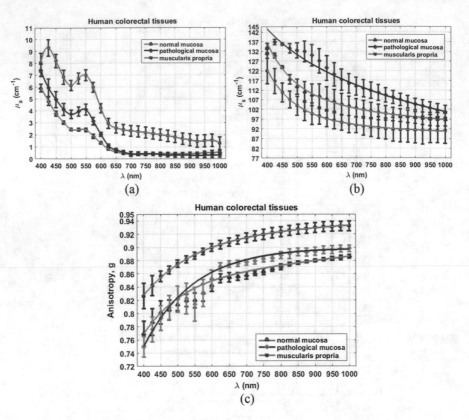

Fig. 1.3 Spectral properties of μ_a (**a**), μ_s (**b**), and g (**c**) for normal mucosa, pathological colorectal mucosa [19], and muscularis propria [20]

$$\mu_{\text{s-nm}}(\lambda) = 115.2\left(0.1161\left(\frac{\lambda}{500\,(\text{nm})}\right)^{-4} + (1 - 0.1161)\left(\frac{\lambda}{500\,(\text{nm})}\right)^{-0.08636}\right),$$

$$(1.10)$$

$$\mu_{\text{s-pm}}(\lambda) = 130.3\left(0.01654\left(\frac{\lambda}{500\,(\text{nm})}\right)^{-4} + (1 - 0.01654)\left(\frac{\lambda}{500\,(\text{nm})}\right)^{-0.3479}\right),$$

$$(1.11)$$

$$\mu_{\text{s-m}}(\lambda) = 104\left(0.1209\left(\frac{\lambda}{500\,(\text{nm})}\right)^{-4} + (1 - 0.1209)\left(\frac{\lambda}{500\,(\text{nm})}\right)^{-0.02265}\right),$$

$$(1.12)$$

$$g_{\text{nm}}(\lambda) = 0.8311 \times e^{6.494 \times 10^{-5}\lambda} - 2.634 \times e^{-8.628 \times 10^{-3}\lambda}, \qquad (1.13)$$

$$g_{pm}(\lambda) = 0.9088 \times e^{9.681 \times 10^{-6}\lambda} - 2.189 \times e^{-6.619 \times 10^{-3}\lambda}, \tag{1.14}$$

$$g_m(\lambda) = 0.9146 \times e^{2.257 \times 10^{-5}\lambda} - 1.163 \times e^{-6.265 \times 10^{-3}\lambda}. \tag{1.15}$$

The wavelength dependence for μ_s is given by Eq. (1.10) for the normal mucosa ($\mu_{s\text{-}nm}$), by Eq. (1.11) for the pathological mucosa ($\mu_{s\text{-}pm}$), and by Eq. (1.12) for the muscle layer of the colorectal wall ($\mu_{s\text{-}m}$). Similarly, the wavelength dependence for g is given by Eq. (1.13) for the normal mucosa (g_{nm}), by Eq. (1.14) for the pathological mucosa (g_{pm}), and by Eq. (1.15) for the muscle layer (g_m).

The data in Fig. 1.3c show some absorption bands approximately at 550 and 750 nm, indicating the presence of blood and lipids for the normal and pathological mucosa. Some evidence of blood accumulation is also seen for pathological mucosa in the graph of Fig. 1.3b. The data for the muscle layer are smooth in the entire spectral range.

At present day, much information is available about μ_s and μ_a for many tissues and blood, and spectra of these properties have been estimated in some cases [1–5, 19, 20, 22–25]. Considering such data for various tissues and blood, five optical windows have been established for diagnosis and therapeutic purposes. Those windows are located at the wavelength ranges: I (350–400 nm), II (625–975 nm), III (1100–1350 nm), IV (1600–1870 nm), and V (2100–2300 nm) [26, 27]. Such spectral ranges have been established due to the existence of local maxima for δ [23]. Other spectral ranges outside these optical windows present reduced penetration depth due to the existence of absorption bands of tissue components.

If we consider the wavelength range between 400 and 1000 nm, water presents low absorption bands, but hemoglobin and lipids have some significant absorption bands within this range (see Fig. 1.2). Additionally to the absorption bands of tissue components, light penetration depth is also small due to the significant difference between the magnitude of μ_s and μ_a. To overcome this problem, it is necessary to understand why μ_s has such huge values when compared to μ_a. Such explanation is presented in the following section.

1.2   Why the Scattering Coefficient Is So High?

As already mentioned, biological tissues are inhomogeneous materials that contain several components, such as collagen fibers, cells, and their organelles [14]. These components are surrounded by tissue fluids, such as the cellular cytoplasm or interstitial fluid (ISF) [1]. Figure 1.4 shows an example of such internal heterogeneous composition.

Figure 1.4 shows a cross section of a skeletal muscle sample, which is a fibrous tissue. Skeletal muscle can be basically described as a distribution of protein fiber cords through the muscle cell cytoplasm (named sarcoplasm) and ISF in-between the

Fig. 1.4 Cross-sectional photograph of a skeletal muscle sample—species *Wistar Han* rat (10× magnification) [28]

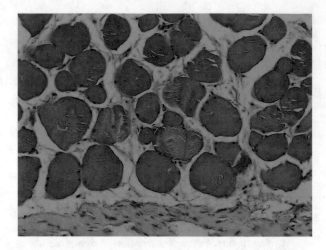

muscle cells [29–31]. This simplified internal description for the muscle indicates that light is strongly scattered inside due to the RI mismatch between sarcoplasm, mitochondria, and protein fibers and between ISF and muscle cells [32].

Since this internal composition of tissues cannot be perceived macroscopically, an optical characterizaton of a tissue as having a mean global refractive index (RI) is many times considered [1]. The mean RI of a tissue (n_{tissue}) can be calculated through Gladstone and Dale equation [14, 33–37], as a weighted sum of the contributions from the various tissue components:

$$n_{\text{tissue}} = \sum_{i=1}^{N} f_i \times n_i; \ \sum_{i=1}^{N} f_i = 1, \tag{1.16}$$

where f_i represents the volume fraction of each tissue component (i) and n_i represents the RI of that component.

When a beam of light travels inside a tissue, the RI values of the individual tissue components become important, since light has to pass from one component to the other. Biological fluids in tissues are composed mainly of water [38], having a small quantity of dissolved salts, proteins, and other organic compounds [33, 39]. Since water is the main component in tissue fluids, its RI will represent the major contribution to the RI of the tissue fluids. The RI of water is known to be significantly lower than the RI of the other tissue components, commonly designated as scatterers. Considering the reference wavelength of an Abbe refractometer (589.6 nm), water has a RI of 1.333 [11], and other tissue components like skin melanin or dry proteins in muscle have RI values of 1.6 [33] and 1.584 [31], respectively. Figure 1.5 presents the dispersions of water and dry proteins in skeletal muscle for the 200–1000 nm range.

Considering the skeletal muscle, the RI of the ISF ranges between 1.35 and 1.38 [33], and the RI of hydrated proteins (scatterers) is 1.53 [41]; the RI mismatch is high as presented in Fig. 1.6.

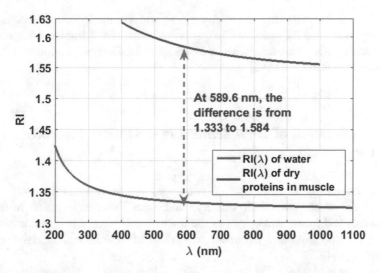

Fig. 1.5 Dispersion of water [14, 40] and dry proteins in skeletal muscle from *Wistar Han* rat [28]

Fig. 1.6 Refractive index mismatch between ISF and scatterers in muscle

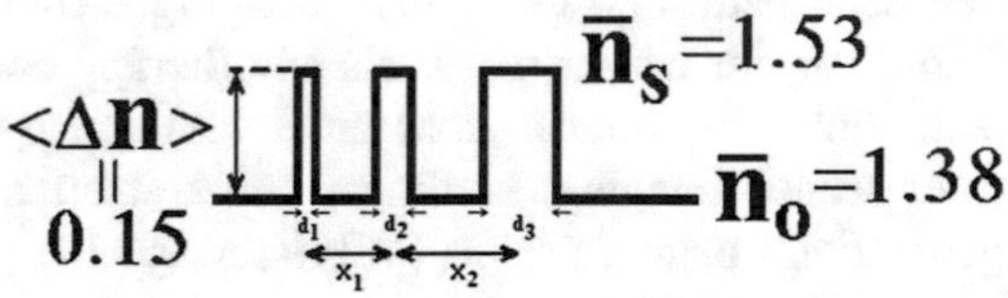

Such difference in RI of tissue components is designated as RI mismatch, and its ratio (m) determines the scattering efficiency of the tissue [33]:

$$m = \frac{n_s}{n_o}. \tag{1.17}$$

Equation (1.17) calculates the scattering efficiency from the RI of the scatterers (n_s) and the RI of ISF (n_o). For a tissue to be completely transparent, m should be 1, i.e., n_o should match n_s. For the case of the muscle considered in the example of Fig. 1.5, $m = 1.14$, meaning that the skeletal muscle has a significantly high μ_s due to the RI mismatch between tissue components. This is the case of various tissues at different wavelengths, as we can see from Table 1.1, Figs. 1.3, and 1.5.

For tissues where the RI of the ISF approaches the RI of scatterers, m decreases to values close to 1. In that case, and according to Eq. (1.18) [42], scattering reduces almost to zero.

$$\mu'_s = \mu_s(1 - g) = 3.28\pi a^2 \rho_s \left(\frac{2\pi \bar{n}_0 a}{\lambda}\right)^{0.37} (m - 1)^{2.09}. \tag{1.18}$$

In Eq. (1.18) the scattering coefficient is defined as the product of the scattering cross section, σ_{sca}, by the volume density of the spheres, ρ_s, as described by

Eq. (1.2), g is the scattering anisotropy factor, and λ is the light wavelength in the scattering medium. Equation (1.18) was validated for samples composed of noninteracting Mie scatterers, when $g > 0.9$; $5 < 2\pi a/\lambda < 50$; $1 < m < 1.1$, what is well fit to many tissues [42].

1.3   Implications of High Scattering in Optical Methods for Clinical Practice

Light is known to have therapeutic properties for many centuries [43]. In the second half of the twentieth century, and with the development of optical components and instruments, such as the laser or the optical fibers, the application of light for diagnosis and treatment of various conditions has received a great breakthrough [43].

Optical technologies offer the possibility of implementing noninvasive or minimally invasive methods for clinical practice. In the last decades, particular interest has been dedicated to in vivo imaging techniques of tissues and cells at high resolution, for cancer diagnosis and therapy control [44]. Some of those imaging techniques are optical coherence tomography (OCT), photoacoustic imaging, fluorescence imaging, nonlinear and Raman microscopy [45], second harmonic generation imaging (SHG) [46], laser speckle imaging [47, 48], and multiphoton microscopy [49, 50].

In addition to imaging methods, there are other optical techniques of great importance in clinical practice. Some examples are terahertz spectroscopy [51], fluorescence spectroscopy [52], or Raman scattering spectroscopy [53]. To improve such methods and develop new ones, the research in Biophotonics has gathered numerous enthusiasts in recent years. Several groups throughout the world perform research in this field, applying optical methods to measure the RI of biological tissues [54–57], studying their optical properties [3–5, 22–25, 58–60], or investigating the efficiency of optical clearing treatments and the diffusion properties of optical clearing agents (OCAs) [61–65].

All the imaging or spectroscopy techniques indicated above and others that are used in Biophotonics applications or related research are limited by the optical properties of biological tissues and blood. Due to the strong scattering in biological materials, light penetration depth is low. In other words, for an imaging or spectroscopic method, such as OCT or reflectance spectroscopy, the acquired images and signals can only be taken from superficial layers of the tissues. If images or signals are to be taken from deeper tissue layers, image contrast or spatial resolution and signal intensity will decrease drastically [1, 66, 67].

To overcome this natural problem of light scattering in biological materials, the optical immersion clearing (hereafter designated as OC) technique was proposed in 1997 [68]. Such technique has been intuitively known from ancient times, since early northern European inhabitants used dried skin to construct the walls of their dwellings or dehydrated seal intestines for windows, so that light could enter inside [69]. Nowadays, this is the basis for technology to estimate the refractive index of paper compounds [70]. The major evidence of the application of this technique for medical purposes was discovered more than 100 years ago by the German anatomist Werner Spalteholz [71, 72], who observed an increase in the transparency of muscle samples that were dehydrated with the use of alcohol, followed by clove oil or xylene, and then stored in Canadian balsam [73–75]. The occurrence of the RI matching mechanism was immediately comprehended by Spalteholz, and the evidence of tissue shrinkage due to initial dehydration mechanism was later reported in 1939 [76]. Since the work of Spalteholz and until the late 1980s, because of no actuality for medical practice, such technique was used only in the preparation of histological specimens. By the early 1990s and as a result of the invention of new lasers and highly efficient light diodes of numerous colors that became available, a new era of light application in biology and medicine was created. As a result of such new instrumentation, new interest to investigate the interaction and propagation of bright light beams in biological tissues, and consequent suppression of light scattering, the OC technique gained new importance. The RI matching method for tissue optical clearing was reinvented to be applied in clinical practice. Since 1997 the OC technique has been thoroughly investigated in combination with other methods, and results show improvement of light penetration depth and scattering reduction in various biological tissues [33, 44, 45, 77].

In Chap. 2, we will introduce this method and present its advantages relative to other methods of reducing light scattering in tissues.

References

1. V.V. Tuchin, *Tissue Optics – Light Scattering Methods and Instruments for Medical Diagnostics*, 3rd edn. (SPIE Press, Bellingham, 2015)
2. T. Vo-Dinh (Ed.). Biomedical Photonics Handbook, Chapter 2, 2nd edition, 1, CRC Press, Boca Raton, 2015
3. A.N. Bashkatov, E.A. Genina, V.V. Tuchin, Optical properties of skin, subcutaneous and muscle tissues: a review. J. Innov. Opt. Health Sci. **4**(1), 9–38 (2011)
4. S.L. Jacques, Optical properties of biological tissues: a review. Phys. Med. Biol. **58**(11), R37–R61 (2013)
5. A.N. Bashkatov, E.A. Genina, V.I. Kochubey, A.A. Gavrilova, S.V. Kapralov, V.A. Grishaev, V.V. Tuchin, Optical properties of human stomach mucosa in the spectral range from 400 to 2000 nm: prognosis for gastroenterology. Med. Laser Appl. **22**(2), 95–104 (2007)
6. I. Yariv, G. Rahamim, E. Shlieselberg, H. Duadi, A. Lipovsky, R. Lubart, D. Fixler, Detecting nanoparticles in tissue using an optical iterative technique. Biomed. Opt. Express **5**(11), 3871–3881 (2014)
7. Y. Zhou, J. Yao, L.V. Wang, Tutorial on photoacoustic tomography. J. Biomed. Opt. **21**(6), 061007 (2016)
8. https://omlc.org/spectra/index.html. Accessed 21 Jan 2019
9. S. Takatani, M.D. Graham, Theoretical analysis of diffuse reflectance from a two-layer tissue model. I.E.E.E. Trans. Biomed. Eng. **BME-26**, 656–664 (1987)

10. R.L. van Veen, H.J. Sterenborg, A. Pifferi, A. Torricelli, E. Chikoidze, R. Cubeddu, Determination of visible near-IR absorption coefficients of mammalian fat using time- and spatially resolved diffuse reflectance and transmission spectroscopy. J. Biomed. Opt. **10**(5), 054004 (2005)
11. G.M. Hale, M.R. Querry, Optical constants of water in the 200nm to 200μm wavelength region. Appl. Opt. **12**(3), 555–563 (1973)
12. V.V. Tuchin, *Lasers and Fiber Optics in Biomedical Science*, 2nd edn. (Saratov University Press, Saratov, Russia, 2010)
13. V.V. Tuchin, S.R. Utz, I.V. Yaroslavsky, Tissue optics, light distribution and spectroscopy. Opt. Eng. **33**(10), 3178–3188 (1994)
14. L. Oliveira, M.I. Carvalho, E. Nogueira, V.V. Tuchin, Skeletal muscle dispersion (400–1000 nm) and kinetics at optical clearing. J. Biophotonics **11**(1), e201700094 (2018)
15. C.T. Germer, A. Roggan, J.P. Ritz, C. Isbert, D. Albrecht, G. Müller, H.J. Buhr, Optical properties of native and coagulated human liver tissue and liver metastases in the near infrared range. Laser. Surg. Med. **23**(4), 194–203 (1998)
16. A.N. Yaroslavsky, P.C. Schultze, I.V. Yaroslavsky, R. Schober, F. Ulrich, H.-J. Schwarzmaier, Optical properties of selected native and coagulated human brain tissues *in vitro* in visible and near infrared spectral range. Phys. Med. Biol. **47**(12), 2059–2073 (2002)
17. W.-F. Cheong, S.A. Prahl, A.J. Welch, A review of the optical properties of biological tissues. IEEE J. Quant. Electron. **26**(12), 2166–2185 (1990)
18. A. Roggan, K. Dörschel, O. Minet, D. Wolff, G. Müller, The optical properties of biological tissue in the near infrared wavelength range – review and measurements, in *Laser-Induced Interstitial Thermotherapy*, ed. by G. Müller, A. Roggan, (SPIE Press, Bellingham, 1995), pp. 10–44
19. S. Carvalho, N. Gueiral, E. Nogueira, R. Henrique, L. Oliveira, V.V. Tuchin, Comparative study of the optical properties of colon mucosa and colon precancerous polyps between 400 and 1000 nm, in *Dynamics and Fluctuations in Biomedical Photonics XIV*, Proc. SPIE, ed. by V. V. Tuchin, K. V. Larin, M. J. Leahy, R. K. Wang, vol. 10063, (SPIE Press, Bellingham, 2017), p. 100631L
20. I. Carneiro, S. Carvalho, R. Henrique, L.M. Oliveira, V.V. Tuchin, Optical properties of colorectal muscle in visible/NIR range, in *Biophotonics: Photonic Solutions for Better Health Care VI*, Proc. SPIE, ed. by J. Popp, V. V. Tuchin, F. S. Pavone, vol. 10685, (SPIE Press, Bellingham, 2018), p. 106853D
21. S.A. Prahl, M.J.C. van Gemert, A.J. Welch, Determining the optical properties of turbid media by using the adding-doubling method. Appl. Opt. **32**(4), 559–568 (1993)
22. A.N. Bashkatov, E.A. Genina, V.I. Kochubey, V.V. Tuchin, Optical properties of human sclera in spectral range 370–2500 nm. Opt. Spectrosc. **109**(2), 197–204 (2010)
23. A.N. Bashkatov, E.A. Genina, M.D. Kozintseva, V.I. Kochubey, S.Y. Gorofkov, V.V. Tuchin, Optical properties of peritoneal biological tissues in the spectral range of 350–2500 nm. Opt. Spectrosc. **120**(1), 1–8 (2016)
24. A.N. Bashkatov, E.A. Genina, V.I. Kochubey, V.S. Rubtsov, E.A. Kolesnikova, V.V. Tuchin, Optical properties of human colon tissues in the 350–2500 spectral range. Quant. Electron. **44**(8), 779–784 (2014)
25. M. Firbank, M. Hiraoka, M. Essenpreis, D.T. Delpy, Measurement of the optical properties of the skull in the wavelength range 650–950 nm. Phys. Med. Biol. **38**(4), 503–510 (1993)
26. D.C. Sordillo, L.A. Sordillo, P.P. Sordillo, L. Shi, R. Alfano, Short wavelength infrared optical windows for evaluations of benign and malignant tissues. J. Biomed. Opt. **22**(4), 045002-1–045002-7 (2017)
27. L. Shi, A. Rodriguez-Contreras, R. Alfano, Transmission in near-infrared optical windows for deep brain imaging. J. Biophotonics **9**(1–2), 38–43 (2016)
28. L. Oliveira, *The effect of optical clearing in the optical properties of skeletal muscle*, PhD thesis, FEUP edições, Porto, Portugal, 2014

29. A.N. Bashkatov, E.A. Genina, V.V. Tuchin, Measurement of glucose diffusion coefficients in human tissues, Chapter 19, in *Handbook of Optical Sensing of Glucose in Biological Fluids and Tissues*, ed. by V. V. Tuchin, (Taylor & Francis Group LLC, CRC Press, London, 2009), pp. 87–621

30. R.C. Haskell, F.D. Carlson, P.S. Blank, Form birefringence of muscle. Biophys. J. **56**, 401–413 (1989)

31. L. Oliveira, A. Lage, M. Pais Clemente, V.V. Tuchin, Optical characterization and composition of abdominal wall muscle from rat. Opt. Laser Eng. **47**(6), 667–672 (2009)

32. V.V. Tuchin, Tissue optics and photonics: biological tissue structures. J. Biomed. Phot. Eng. **1**(1), 3–21 (2015)

33. V.V. Tuchin, *Optical Clearing of Tissues and Blood* (SPIE Press, Bellingham, 2006)

34. D.W. Leonard, K.M. Meek, Refractive indices of the collagen fibrils and extracellular material of the corneal stroma. Biophys. J. **72**(3), 1382–1387 (1997)

35. K.M. Meek, S. Dennis, S. Khan, Changes in the refractive index of the stroma and its extrafibrillar matrix when the cornea swells. Biophys. J. **85**(4), 2205–2212 (2003)

36. K.M. Meek, D.W. Leonard, C.J. Connon, S. Dennis, S. Khan, Transparency, swelling and scarring in the corneal stroma. Eye **17**(8), 927–936 (2003)

37. O. Zhernovaya, O. Sydoruk, V.V. Tuchin, A. Douplik, The refractive index of human hemoglobin in the visible range. Phys. Med. Biol. **56**(13), 4013–4021 (2011)

38. R.F. Reinoso, B.A. Telfer, M. Rowland, Tissue water content in rats measured by desiccation. J. Pharmacol. Toxicol. Methods **38**(2), 87–92 (1997)

39. L. Oliveira, A. Lage, M. Pais Clemente, V.V. Tuchin, Rat muscle opacity decrease due to the osmosis of a simple mixture. J. Biomed. Opt. **15**(5), 055004-1–055004-9 (2010)

40. M. Daimon, A. Masumura, Measurement of the refractive index of distilled water from the near-infrared region to the ultraviolet region. Appl. Opt. **46**, 3811–3820 (2007)

41. W.L. Bragg, A.B. Pippard, The form birefringence of macromolecules. Acta Cryst **6**, 865–867 (1953)

42. R. Graaff, J.G. Aarnoudse, J.R. Zijp, P.M.A. Sloot, F.F. de Mul, J. Greve, M.H. Koelink, Reduced light-scattering properties for mixtures of spherical particles: a simple approximation derived from Mie calculations. Appl. Opt. **31**(10), 1370–1376 (1992)

43. R. Splinter, B.A. Hooper, *An Introduction to Biomedical Optics* (Taylor and Francis, New York, 2007)

44. A.Y. Sdobnov, M.E. Darvin, E.A. Genina, A.N. Bashkatov, J. Lademann, V.V. Tuchin, Recent progress in tissue clearing for spectroscopic application. Spectrochim. Acta A Mol Biomol. Spectrosc. **197**, 216–229 (2018)

45. E.A. Genina, A.N. Bashkatov, Y.P. Sinichkin, I.Y. Yanina, V.V. Tuchin, Optical clearing of biological tissues: prospects of application in medical diagnosis and phototherapy. J. Biomed. Photon. Eng. **1**(1), 22–58 (2015)

46. F. S. Pavone, P. J. Campagnola (eds.), *Second Harmonic Generation Imaging* (CRC Press, Boca Raton, 2014)

47. V. V. Tuchin (Ed.). Handbook of Optical Biomedical Diagnostics. 2nd ed., vols. 1 & 2, SPIE Press, Bellingham, 2016

48. A.F. Fercher, J.D. Briers, Flow visualization by means of single-exposure speckle photography. Opt. Commun. **37**(5), 326–330 (1981)

49. M.E. Darvin, H. Richter, Y.J. Zhu, M.C. Meinke, F. Knorr, S.A. Gonchukov, K. Koenig, J. Lademann, Comparison of *in vivo* and *ex vivo* laser scanning microscopy and multiphoton tomography application for human and porcine skin imaging. Quant. Electron. **44**(7), 646–651 (2014)

50. M. Ulricht, M. Klemp, M.E. Darvin, K. Konig, J. Lademann, M.C. Meinke, In vivo detection of basal cell carcinoma: comparison of a reflectance confocal microscope and a multiphoton tomograph. J. Biomed. Opt. **18**(6), 061229 (2013)

51. O.A. Smolyanskaya, I.J. Schelkanova, M.S. Kulya, E.L. Odlyanitskiy, I.S. Goryachev, A.N. Tcypkin, Y.V. Grachev, Y.G. Toropova, V.V. Tuchin, Glycerol dehydration of native and diabetic animal tissues studied by THz-TDS and NMR methods. Biomed. Opt. Express **9**(3), 1198–1215 (2018)

52. L. Lim, B. Nichols, N. Rajaram, J.W. Tunnell, Probe pressure effects on human skin diffuse reflectance and fluorescence spectroscopy measurements. J. Biomed. Opt. **16**(1), 011012 (2011)

53. B. Broadbent, J. Tseng, R. Kast, T. Noh, M. Brusatori, S.N. Kalkanis, G.W. Auner, Shining light on neurosurgery diagnostics using Raman spectroscopy. J. Neurooncol. **130**(1), 1–9 (2016)

54. Z. Deng, J. wang, Q. Ye, T. Sun, W. Zhou, J. Mei, C. Zhang, J. Tian, Determination of continuous complex refractive dispersion of biotissue based on internal reflection. J. Biomed. Opt. **21**(1), 015003 (2016)

55. H. Ding, J.Q. Lu, W.A. Wooden, P.J. Kragel, X.-H. Hu, Refractive indices of human skin tissues at eight wavelengths and estimated dispersion relations between 300 and 1600 nm. Phys. Med. Biol. **51**(6), 1479–1489 (2006)

56. H. Li, S. Xie, Measurement method of the refractive index of biotissue by total internal reflection. Appl. Opt. **35**(10), 1793–1795 (1996)

57. I. Carneiro, S. Carvalho, R. Henrique, L. Oliveira, V.V. Tuchin, Water content and scatterers dispersion evaluation in colorectal tissues. J. Biomed. Phot. Eng. **3**(4), 040301-1–040301-10 (2017)

58. A. Vogel, C. Dlugos, R. Nuffer, R. Birngruber, Optical properties of human sclera, and their consequences for transscleral laser applications. Laser. Surg. Med. **11**(4), 331–340 (1991)

59. A. Roggan, M. Friebel, K. Dörschel, A. Hahn, G. Müller, Optical properties of circulating human blood in the wavelength range 400–2500 nm. J. Biomed. Opt. **4**(1), 36–46 (1999)

60. A.N. Bashkatov, E.A. Genina, V.I. Kochubey, V.V. Tuchin, Optical properties of human skin, subcutaneous and mucous tissues in the wavelength range from 400 to 2000 nm. J. Phys. D. Appl. Phys. **38**(15), 2543–2555 (2005)

61. D.K. Tuchina, R. Shi, A.N. Bashkatov, E.A. Genina, D. Zhu, V.V. Tuchin, Ex vivo optical measurements of glucose diffusion kinetics in native and diabetic mouse skin. J. Biophotonics **8**(4), 332–346 (2015)

62. D.K. Tuchina, A.N. Bashkatov, A.B. Bucharskaya, E.A. Genina, V.V. Tuchin, Study of glycerol diffusion in skin and myocardium ex vivo under the conditions of developing alloxan-induced diabetes. J. Biomed. Phot. Eng. **3**(2), 020302 (2017)

63. S. Carvalho, N. Gueiral, E. Nogueira, R. Henrique, L. Oliveira, V.V. Tuchin, Glucose diffusion in colorectal mucosa: a comparative study between normal and cancer tissues. J. Biomed. Opt. **22**(9), 091506 (2017)

64. L. Oliveira, M.I. Carvalho, E. Nogueira, V.V. Tuchin, Diffusion characteristics of ethylene glycol in skeletal muscle. J. Biomed. Opt. **20**(5), 051019 (2015)

65. L. Oliveira, M.I. Carvalho, E. Nogueira, V.V. Tuchin, The characteristic time of glucose diffusion measured for muscle tissue at optical clearing. Laser Phys. **23**(7), 075606 (2013)

66. D.A. Boas, A fundamental limitation of linearized algorithms for diffuse optical tomography. Opt. Express **1**(13), 404–413 (1997)

67. C.L. Smithpeter, A.K. Dunn, A.J. Welch, R. Richards-Kortum, Penetration depth limits of in vivo confocal reflectance imaging. Appl. Opt. **37**(13), 2749–2754 (1998)

68. V.V. Tuchin, I.L. Maksimova, D.A. Zimnyakov, I.L. Kon, A.H. Mavlutov, A.A. Mishin, Light propagation in tissues with controlled optical properties. J. Biomed. Opt. **2**(4), 401–417 (1997)

69. I. Carneiro, S. Carvalho, R. Henrique, L. Oliveira, V.V. Tuchin, Moving tissue spectral window to the deep-UV via optical clearing. J. Biophot. (2019). https://doi.org/10.1002/jbio.201900181

70. T. Fabritius, E. Alarousu, T. Prykäri, J. Hast, R. Myllylä, Characterization of optically cleared paper by optical coherence tomography. Quant. Electron. **36**(2), 181–187 (2006)

71. W. Spalteholz, *Über das Durchsichtigmachen von menschlichen unde tierischen Präparaten und seine theoretichen Bedingungen, nebst Anhang: Über Knochenfärbung* (S. Hirszel, Leipzig, Germany, 1911)

72. W. Spalteholz, *Über das Durchsichtigmachen von menschlichen unde tierischen Präparaten und seine theoretichen Bedingungen, nebst Anhang: Über Knochenfärbung* (S. Hirszel, Leipzig, Germany, 1914)
73. D.S. Richardson, J.W. Lichtman, Clarifying tissue clearing. Cell **162**(2), 246–257 (2015)
74. M. Aswendt, M. Schwarz, W.M. Abdelmoula, J. Dijkstra, S. Dedeurwaerdere, Whole-brain microscopy meets *in vivo* neuroimaging: techniques, benefits, and limitations. Mol. Imaging Biol. **19**(1), 1–9 (2017)
75. A. Azaripour, T. Lagerweij, C. Scharfbillig, A.E. Jadczac, B. Willershausen, C.J. Van Noorden, A survey of clearing techniques for 3D imaging of tissues with special reference to connective. Prog. Histochem. Cytochem. **51**(2), 9–23 (2016)
76. R.W. Cumley, J.F. Crow, A.B. Griffen, Clearing specimens for the demonstration of bone. Biotech. Histochem. **14**, 7–11 (1939)
77. E.A. Genina, A.N. Bashkatov, V.V. Tuchin, Tissue optical immersion clearing. Expert Rev. Med. Dev. **7**(6), 825–842 (2010)

Chapter 2
Controlling the Optical Properties of Biological Materials

2.1 Turning the Tissues Clear: An Introduction

One of the greatest objectives in Biophotonics is to turn biological tissues clear and allow for improved optical technologies in clinical practice. For several years different and alternative methods are known to be capable to reduce light scattering in tissues. Such methods are diverse, and they present some advantages and downsides when compared between each other. Increase of tissue turbidity, on the other hand, may also be of interest for imaging purposes. With the objective of analyzing the various possibilities to improve the optical methods in clinical practice, the following sections are used to make a description of the different methods to reduce or increase light scattering.

2.2 Tissue Whitening

Optical imaging techniques have been used for long time to perform diagnosis. Within the imaging methods, light-sheet microscopy [1], multiphoton microscopy, confocal microscopy, Raman microscopy, optical coherence tomography (OCT), and speckle imaging are the most recent [2].

For any imaging method to establish a reliable diagnosis, it is necessary that high contrast is obtained in the acquired images of tissues or cells [2]. Since for the majority of applications, images are obtained from reflected light, one way to increase image contrast of transparent tissues like human cervical mucosa is to induce tissue turbidity instead of making the tissue clear. This process is designated as tissue whitening. Studies have demonstrated that the application of acetic acid to tissues enhances contrast in confocal images of in vivo tissues, allowing one to locate dysplastic areas [3–6]. This method has been successfully applied to study

© The Author(s), under exclusive license to Springer Nature Switzerland AG 2019
L. M. C. Oliveira, V. V. Tuchin, *The Optical Clearing Method*,
SpringerBriefs in Physics, https://doi.org/10.1007/978-3-030-33055-2_2

neoplasia or precancerous lesions in various anatomical locations, such as the larynx [6], Barrett's esophagus [7], or the cervix [8, 9].

Topical application of low-concentrated acetic acid solutions (3–6%) is commonly used in this method. As observed by some authors [3, 6], the application of acetic acid into cervix and larynx induces transient whitening changes in the epithelial tissues [10]. The differential contrast between normal and dysplastic areas in the tissue allows for a diagnosis. Such transient effect, designated as acetowhitening, is fast, occurring within the first 2–3 min [3], and normally decays within 5–10 min [5]. The use of this method in the cervix has been made for several decades, and no side effects have been reported [6]. As alternative to acetic acid application, toluidine blue or photosensitizers can produce an increase in image contrast or in fluorescence signal, but the process takes several hours to occur, and toluidine blue is known to produce false positive [6].

Some research has been carried out to study the mechanisms associated with acetowhitening [9, 11, 12], but the fundamental kinetic mechanism involved in this process at the subcellular level remains to be explained in detail [11].

Some hypotheses have been formulated. One says that acetic acid somehow induces protein structure changes, which leads to increased light scattering which due to bigger optical path for photons underlines absorption properties in normally transparent tissues and allows for detection of weakly absorbing lesions in the epithelium. The mechanism for structure changes in proteins is not exactly explained, but one explanation has been proposed. It says that in normal conditions, cell nucleus contains a diffuse network of thin chromatin filaments with diameter ranging typically between 30 and 100 nm. These filaments have a RI within the range 1.38–1.41 [13] and occupy a small volume. Cell cytoplasm and extracellular fluid have a RI between 1.350 and 1.375 [13]. Due to the low RI mismatch, light backscattering is low. Acetic acid application causes the chromatin fibers to assemble together into thick fibers with diameters of 1–5 μm that will fill a large fraction of the intranuclear volume and increase the RI inside the nucleus [4, 10]. These changes will increase the backscattering signal from the nuclei making them to appear bright. Again, since the dysplastic areas have more concentration of proteins, the effect will be higher, and the backscattering signal will be brighter than in normal areas of the tissue. Consequently, and since the dysplastic areas have higher protein concentration than normal areas, the acetowhitening effect is higher in these areas, allowing for differentiation [4].

The use of polarized light and image processing techniques in confocal microscopy has allowed to make significant improvements in this technique, and high-contrast images have been obtained [6].

2.3　Temperature Effects and Tissue Coagulation

Temperature variations are known to produce changes in the optical properties of biological tissues in different ways. Such changes are diverse, and a detailed explanation of the various temperature-related effects may be found in literature.

In the present section, we will discuss some temperature-related effects that are known to change the optical properties of biological tissues and the corresponding benefits and downsides in clinical practice.

Temperature is indeed an important factor for some of tissue components, since as we can see from Fig. 1.2, among other chromophores, tissues have water, lipids, and blood hemoglobin. These three major tissue components have distinct absorption bands at particular wavelengths within the visible and near-infrared (NIR) wavelength ranges. By heating the tissues, water and lipid molecules, and also proteins may disrupt, leading to a decrease in the corresponding absorption bands.

One particularly interesting effect consists on a wavelength shift of the absorption bands, as a consequence of temperature variations [14]. Water absorption peaks at the NIR are of particular interest for laser surgery, since at those wavelengths (1.94 and 2.95 µm) precise tissue ablation can be performed due to limited light penetration depth [14]. Studies [15] have showed that absorption of albumin near 1.94 µm is due to the presence of water and that this local maximum in the absorption coefficient shifts to shorter wavelengths with increasing temperature [14]. Experimental results indicate that similar absorption shift may occur in other biological materials [14, 16]. A study was conducted with a tunable (1.88–2.07 µm) Tm:YAG laser to analyze the necessary irradiance to create a coagulum in albumin samples at different temperatures [14]. From this study, it was observed that for any particular wavelength near 1.94 µm, as the temperature rises from 21.6 to 28.0 °C and then to 41.6 °C, the necessary laser irradiance to create the lesion within a constant exposure time decreases with increasing temperature. For the samples heated at 41.6 °C, the minimum necessary irradiance has been observed for 1.92 µm, indicating that a water absorption shift has occurred [14, 15].

The increase of blood temperature may also affect the absorption bands of hemoglobin, both in magnitude and in shape [17–21]. Contradictory behavior has been observed for the transmission of probe wavelengths through 200 µm thick samples of diluted whole blood at a hematocrit (Hct) of 24.5% prior to coagulation and in μ_a, when heated up to 76 °C with a 10 ms pulse of 532 nm laser [22]. Table 2.1 shows the variations obtained in this study.

The data in Table 2.1 was obtained assuming fully oxygenated blood with 18 g H_b/l. The μ_a of blood at room temperature is indicated as μ_{aRT} in Table 2.1. The % change in μ_a over the indicated temperature range is represented as $\Delta\mu_a$, while $\Delta\mu_a(T)$ represents the change in the μ_a per °C over the same temperature range.

Table 2.1 Changes in the μ_a of diluted blood (24.5% Hct) heated from 22 to 76 °C [14]

Wavelength (nm)	μ_{aRT} (cm^{-1})	$\Delta \mu_a$ 22–76 °C (%)	$\Delta \mu_a(T)$ (cm^{-1}/°C)	Δ Transmittance 22–76 °C (%)
532	123.0	−4.0	−0.093	20.0
594	23.90	28.8	0.118	−50.0
612	3.74	15.5	0.0113	−17.9
633	1.39	9.6	0.0027	−5.3
675	1.10	0	0.00	0

The last column of Table 2.1 presents the observed changes in sample transmittance due to the changes in μ_a [22, 23]. These variations show a wavelength shift in the absorption bands of hemoglobin, since for 532 nm, we see a decrease in μ_a, while for 594 nm, we see an increase. Similar behavior was observed when the samples were pumped with a 10 ms, 1064 nm laser, showing that such wavelength shift is a photothermal effect [22]. These temperature changes in μ_a should be of high importance when planning vascular surgery, so that appropriate laser wavelength and light dosage are properly selected [24, 25].

Another important temperature effect is protein denaturation. One study [26] has showed that placing human aorta samples in a water bath (100 °C) for 300 s leads to an increase in the μ_s' of the samples of 10–45% in the visible range and over 150% in the NIR range. A different study with various bovine tissues has demonstrated that heating of samples between approximately 30 and 80 °C leads to an increase in μ_s' at 633 nm for bovine liver and bovine muscle, while brain matter keeps almost unchanged [27]. The increase in μ_s' for liver and muscle was also observed in this study at 810 nm. Although the increase in tissue scattering may be interesting for certain applications like in vivo imaging, it is not beneficial for applications that use light transmission or to increase light penetration depth.

Thermal heating of tissues may also lead to water evaporation. Biological tissues contain water both in the interstitial locations and inside the cells. The interstitial water is the first to evaporate when the tissue is heated. Such water loss in the interstitial locations leads to an approach between the other tissue components (scatterers) and a decrease in tissue thickness [14, 28]. This approach between scatterers leads to an increase in the scattering coefficient, but due to a less sample thickness and better scatterer ordering (packing), light transmittance increases [28]. Since according to Gladstone and Dale equation (Eq. 1.9), the RI of the tissue is a combination between the RIs of tissue components, water evaporation leads also to a change in the RI of the tissue. Water loss leads to a decrease in its volume fraction and an increase of the volume fractions of the other tissue components. Consequently, since, in general, other tissue components have a higher RI than water, the RI of the tissue will increase [28].

Lowering tissue temperature also leads to a change in the optical properties. Studies performed on eye tissues have reported the occurrence of the so-called cold cataract when temperature decreases [29, 30]. This effect is created due to protein aggregation and translates an increase in the scattering coefficient. Such effect can be reversed by increasing tissue temperature.

Cryosurgery is a clinical procedure that uses extreme cold temperatures to destroy or remove diseased tissue. During the freezing process at such low temperatures (−80 °C), water ice may be created in certain locations inside the tissue. The additional RI variations at the boundary between liquid and frozen water lead to an increase in the scattering coefficient of the tissue [31]. Such increase may be beneficial, if OCT imaging is used to monitor the surgery. It has been observed for hamster skin that such change in tissue scattering allowed seeing the subsurface morphological changes [31].

2.4 Mechanical Tissue Compression and Stretching

The optical properties of biological tissues can also be controlled due to the application of mechanical forces, such as compression and stretching. These mechanical forces remove the interstitial water and blood from the area of interest, leading to an increase of optical tissue homogeneity. Due to the interstitial water loss, closer packing of tissue components causes less light scattering due to cooperative interference effects and thinner tissue [32].

Tissues that contain little blood, such as sclera, however, are characterized by some inertia for a few minutes when submitted to these mechanical forces. This fact is due to the relatively slow water diffusion from the compressed region [33]. An explanation for the mechanical clearing of the sclera has been proposed. Due to sclera compression, water is displaced from the interstitial space between collagen fibrils, leading to an increase of the protein and mucopolysaccharide concentrations. Since these proteins and sugars have a RI more approximated to that of the collagen fibrils, a RI matching occurs in the compressed region. At the same time, mechanical compression reduces sclera thickness, d, which increases the effective scatterer concentration inside [34]. Consequently, compression might also increase μ_s for the sclera. Since the global change in the optical properties of the tissue is proportional to $\mu_s \times d$, less scattering of light occurs.

During tissue compression or stretching, the increase of scatterer concentration can sometimes be more intense than the reduction in index mismatch during tissue compression or stretching [34]. For bloodless tissues or tissues having aggregated or coagulated blood, thickness reduction causes an increase in local chromophore concentration. This means an increase in μ_a [32].

By performing studies in ex vivo sclera samples, the authors of Ref. [34] have observed that compression caused leaking of extracellular fluids from the samples. Intracellular fluids, on the other hand, were kept inside, since an excessive pressure is needed to rupture the cell walls. Due to sample thickness decrease, the volume fraction of intracellular water has increased, which may explain the increase in μ_a at the wavelengths of water bands with compression [34].

Other compression studies have been performed with ex vivo samples of human and porcine skin [35]. In these studies, the IAD method was used to estimate the μ_a and μ_s' of the tissues from spectra measured with integrating sphere in the range from 400 to 1800 nm. Diffuse reflectance and transmittance spectra were measured from tissue samples with an area of 2×2 cm^2 at zero pressure and at external pressures of 0.1, 1, and 2 kg/cm^2 uniformly distributed over the sample surface. A decrease in reflectance was generally observed, while transmittance, μ_a, and μ_s' were increased due to compression. Table 2.2 contains some of the data relative to porcine skin.

As already explained in Chap. 1, the amount of scattering in a tissue depends on the RI mismatch. As we see from the above explanation, it also depends on the scatterer concentration and spacing.

As mechanical forces are applied to biological tissues, the spacing between tissue components is reduced, and water is forced to move outward the compressed or

Table 2.2 Comparative results for ex vivo porcine skin at air dehydration and mechanical compression [35]

	Air dehydration		Mechanical compression	
	Before	After	Before	After
Thickness (μm)	1700 ± 140	680 ± 220	1300 ± 100	540 ± 150
Strain = $\Delta d/d_0$ (%)	–	−59.8 ± −9.6	–	−58.5 ± −8.3
Refractive index	1.36 ± 0.02	1.49 ± 0.03	1.39 ± 0.02	1.5 ± 0.05
Water volume fraction	0.68 ± 0.09	0.35 ± 0.12	0.66 ± 0.02	0.20 ± 0.05
Water weight fraction	0.57 ± 0.01	0.37 ± 0.01	–	–

Error shown as ±1 SD

stretched area. The combination of these two effects leads to a RI matching, causing a decrease in the average light scattering. Ultimately, an increase in tissue transmittance and a decrease in tissue reflectance are observed. On the other hand, sample thickness reduction as a consequence of tissue compression or stretching also leads to an increase in the effective concentration of scatterers and chromophores inside the tissue. This concentration increase causes a certain elevation in μ_s and μ_a. However, the transmitted light intensity should increase, since certain elevations of these coefficients do not strongly affect this product $(\mu_s + \mu_a) \times d$ because thickness decrease at compression can be changed significantly (up to 50–70%) [35, 36].

2.5 The Immersion Method

We have seen in the previous sections that the various alternative methods to control the optical properties of biological tissues present certain advantages and downsides.

These methods are useful, within certain limitations, for a certain number of particular applications.

The optical immersion of tissues in exogenous chemical agents is the most recent and with most potential technique to control the optical properties of biological tissues. As referred in Chap. 1, the first evidence that this method could be used for medical purposes was obtained by Spalteholz in the early 1900s [37, 38]. However, only in the late 1980s and early 1990s, this technology was started to be purposefully developed to solve the problems of rapidly developing optical methods of medical diagnostics and therapy [10, 39, 40]. Much has been done in the meantime [2, 28, 41].

A group of biocompatible agents is known at present day that produce a completely reversible clearing effect in various soft biological tissues or fluids in the visible-NIR range through the reduction of light scattering [10]. The temporary clearing effect created in tissues by these agents produces an increase of light penetration depth and contrast of the acquired images of different imaging methods [2]. With research groups distributed worldwide, this method has produced in the past two decades a large number of publications, and the numbers are still growing [2, 41, 42]. In the following chapters, we will describe the mechanisms behind the immersion method and the measurements that can be performed during treatments along with the valuable information they contain and present the improvements that this method brings to other optical clinical applications.

The immersion method can be used to control the optical properties of biological materials on ex vivo or in vivo situations. For in vivo application, after the tissues are cleared, water from adjacent tissues will reverse the effect, while for the ex vivo situation, the effect can be reversed with assisted rehydration. To make a brief explanation of the optical immersion technique, we will consider ex vivo skeletal muscle samples with a slab-form as an example.

Skeletal muscle is a fibrous tissue that contains long dielectric cylinders (actin and myosin fibers), which are grouped in fiber cords with an average diameter of 53 μm [43] and RI $n_c = 1.53$ (at 589.6 nm) [44]. The muscle fiber cords are located in planes that are parallel to the slab surface but with random orientations, as demonstrated by the different cross sections seen in Fig. 1.4. The space between the fiber cords is filled with the interstitial fluid (ISF), which is composed mainly from water and some dissolved salts and proteins [45]. For skeletal muscle, the RI of the ISF at 589.6 nm, n_0, usually ranges between 1.35 and 1.37 [28]. The considerable mismatch between the RIs of the fiber cords and of the ISF makes the muscle turbid, meaning that it causes multiple scattering and poor transmittance of light [10]. The RI of the ISF can be controlled, and it may be increased to approximate or match the RI of the fiber cords. Such elevation in the RI of the ISF can be made by replacing the interstitial water by an agent with a higher RI. This change turns the tissue from multiple to low-step or even to single-step scattering mode. Ideally, if the absorption is small and when $n_c = n_0 = 1.53$, the muscle becomes totally homogeneous and optically transparent [10].

The model described for the muscle can be used to characterize other fibrous tissues, such as the cornea or the skin dermis, if the appropriate RIs, fiber diameter, and their spacing are provided.

The diffusion of an agent into the interstitial locations of the muscle can be studied if we consider collimated transmission measurements made from a slab-form muscle sample under immersion treatment with an aqueous solution containing the active agent. Such slab-form muscle sample is schematically represented in Fig. 2.1, along with the geometry to measure light transmission of a collimated wideband light beam.

The collimated transmittance of such a slab tissue with thickness d, as represented in Fig. 2.1, is described by Bouguer-Beer-Lambert law [10]:

$$T_c(\lambda, d) = I(\lambda, d)/I(\lambda, 0) = \exp\left[-\mu_t(\lambda) \times d\right], \tag{2.1}$$

where $I(\lambda, 0)$ and $I(\lambda, d)$ are the intensities of the incident and detected light and $\mu_t(\lambda) = \mu_a(\lambda) + \mu_s(\lambda)$ is the attenuation coefficient at wavelength λ. As represented in Table 2.3, μ_s for the rat muscle is much higher than μ_a for the wavelength range between 400 and 1000 nm.

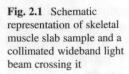

Fig. 2.1 Schematic representation of skeletal muscle slab sample and a collimated wideband light beam crossing it

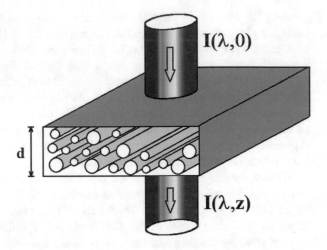

Due to the fibrous structure of the skeletal muscle, it is reasonable to consider that the kinetics of fluid diffusion inside the tissue can be approximated using model of free diffusion with a modified (effective) diffusion coefficient [46–48]. When a chemical agent diffuses into the interfibrillar space of the muscle, we can use the free diffusion model with the approximate solution of the diffusion equation to describe the RI kinetics and corresponding decrease in the scattering coefficient [46, 48]:

$$\frac{\partial C_a(x,t)}{\partial t} = D_a \frac{\partial^2 C_a(x,t)}{\partial x^2}, \qquad (2.2)$$

where $C_a(x,t)$ is the agent concentration, D_a is the diffusion coefficient, and x is the spatial coordinate in the perpendicular axis to the slab surface. This equation is valid when the process rate is not limited by membranes [32, 48].

If we consider a slab tissue sample (see Fig. 2.1) with thickness d, which is immersed in a solution with initial concentration of the agent C_{a0}, we can describe the time dependence for the agent concentration distribution inside the sample. At $t = 0$ the initial concentration of agent inside the tissue equals 0: $C_a(x,0) = 0$ for $t = 0$ and $0 \leq x \leq d$. The boundary conditions are $C_a(0,t) = C_a(d,t) = C_{a0}$. To describe the agent concentration within the sample, Eq. (2.2) has the following solution [32, 46, 48, 49]:

$$C_a(x,t) = C_{a0}\left\{1 - \frac{4}{\pi}\left[\exp\left(-\frac{t}{\tau}\right)\sin\left(\frac{\pi x}{d}\right) + \frac{1}{3}\exp\left(-\frac{9t}{\tau}\right)\sin\left(\frac{3\pi x}{d}\right)\right.\right.$$
$$\left.\left. + \frac{1}{5}\exp\left(-\frac{25t}{\tau}\right)\sin\left(\frac{5\pi x}{d}\right) + \dots\right]\right\}, \qquad (2.3)$$

where

Table 2.3 Optical properties of rat skeletal muscle between 400 and 1000 nm [28]

λ (nm)	μ_a (cm^{-1})	μ_s (cm^{-1})	μ_s' (cm^{-1})	g
400	7.2	119.1	21.2	0.822
425	5.8	114.6	19.6	0.829
450	4.9	111.0	18.3	0.835
475	4.4	108.2	17.3	0.840
500	4.1	105.9	16.4	0.845
525	3.9	104.0	15.6	0.850
550	3.4	102.4	15.0	0.854
575	2.8	101.1	14.4	0.857
600	2.1	99.9	13.9	0.861
625	1.5	98.9	13.5	0.864
650	1.3	98.0	13.1	0.866
675	1.1	97.2	12.7	0.869
700	1.1	96.5	12.4	0.872
725	1.1	95.8	12.1	0.874
750	1.1	95.2	11.8	0.876
775	1.1	94.6	11.5	0.878
800	1.1	94.1	11.3	0.880
825	1.2	93.6	11.1	0.882
850	1.2	93.2	10.9	0.884
875	1.2	92.7	10.6	0.885
900	1.1	92.3	10.5	0.887
925	1.2	91.9	10.3	0.888
950	1.2	91.6	10.1	0.890
975	1.2	91.2	9.9	0.891
1000	1.2	90.9	9.8	0.892

$$\tau = \frac{d^2}{\pi^2 D_a} \tag{2.4}$$

is the diffusion time and D_a is the diffusion coefficient of the agent into the tissue.

At a particular time t of treatment, the ratio of the amount of agent in the tissue, m_t, to its equilibrium value, m_{eq}, is defined as [48]:

$$\frac{m_t}{m_{eq}} = \frac{\int_0^d C_a(x,t)dx}{C_{a0}d}$$
$$= 1 - \frac{8}{\pi^2}\left\{\exp\left(-\frac{t}{\tau}\right) + \frac{1}{9}\exp\left(-\frac{9t}{\tau}\right) + \frac{1}{25}\exp\left(-\frac{25t}{\tau}\right) + \ldots\right\}. \tag{2.5}$$

This ratio represents the volume-averaged concentration of the agent in the tissue, $C_a(t)$, which for a first-order approximation can be represented as [32, 44, 45, 50]:

$$C_a(t) = \frac{1}{2} \int_0^d C_a(x, t) dx \cong C_{a0} \left[1 - \exp\left(-\frac{t}{\tau}\right) \right].$$ (2.6)

If a slab tissue sample is immersed in a solution containing a low molecular weight chemical agent with known diffusion coefficient, D_a, Eqs. (2.3)–(2.6) allow for determining the time-dependent concentration of the agent at depth x or variations of the total amount of the agent within the tissue as it flows in. Conversely, measurement of $C_a(t)$ makes it possible to estimate the D_a value of the agent molecules in the interstitial space of the tissue.

This is the case when T_c spectra are measured from an ex vivo tissue sample during an immersion treatment. T_c data obtained during treatment is sensitive to changes in the interstitial space of the tissue sample, and a different form of Eq. (2.6) can be used to estimate the diffusion time of the agent [51–54]:

$$T_c(\lambda, t) = \frac{C_a(t)}{C_{a0}} \cong \left[1 - \exp\left(-\frac{t}{\tau}\right) \right].$$ (2.7)

Depending on the agent concentration in the treating solution, those measurements translate a diffusion time, τ, for a mixed flux. The osmotic pressure created on the tissue by the agent in the treating solution forces tissue dehydration [53]. This way, the water in the ISF flows out, and the agent molecules flow into the ISF. The value for τ that is estimated by fitting T_c data with Eq. (2.7) represents the diffusion time of such mixed flux. In Chap. 5 we will describe in detail a method for accurate discrimination of the diffusion time values for water flux out of the tissue and agent flux into the tissue and will present some results obtained with this method that allowed to discriminate between normal and pathological tissues.

If through Eq. (2.7) we can estimate τ for the water and agent fluxes during OC, we can use Eq. (2.4) to calculate the corresponding diffusion coefficients. If, on the other hand, the diffusion of the agent is made only through one sample surface, like, for instance, on in vivo topical application of creams and lotions on the skin, all previous equations are still valid, but Eq. (2.4) takes the following form [49]:

$$\tau = \frac{4d^2}{\pi^2 D_a}.$$ (2.8)

Considering low molecular weight compounds, the values of their diffusion coefficients in their own media are approximately 10^{-5} cm²/s: water— $D_a = 2.5 \times 10^{-5}$ cm²/s [48] and glycerol—$D_a = 0.6 \times 10^{-5}$ cm²/s [55]. In reality, when OC treatments are applied to tissues, both the water flux out and the agent flux in occur simultaneously at least during some extent of the treatment. For this reason, the diffusion coefficients of OCAs in water and for water in some OCAs are also important and more approximated to tissue reality. Studies have been made to evaluate mutual diffusion coefficients, which are temperature-dependent [56–58]. Table 2.4 presents some of these mutual diffusion coefficients for water + ethylene glycol and water + glycerol solutions.

Table 2.4 Mutual diffusion coefficients for the systems water (1) + ethylene glycol (2) and water (1) + glycerol (2) at 25 °C [58]

w_2	Δw_2	$10^6 \, D_{12}/\text{cm}^2/\text{s}$ Min	Max	Mean	$10^6 \, \sigma/\text{cm}^2/\text{s}$	No. of data points
Mass fraction of ethylene glycol						
0.025	0.0500	11.02	12.30	11.70	0.32	3
0.200	0.0800	9.22	9.82	9.58	0.16	3
0.400	0.0800	7.33	8.00	7.64	0.19	2
0.600	0.0800	5.65	6.26	5.95	0.14	3
0.800	0.0800	4.11	4.94	4.63	0.19	3
0.950	0.100	3.47	4.08	3.75	0.15	3
Mass fraction of glycerol						
0.020	0.0400	9.28	10.23	9.77	0.30	4
0.200	0.0400	7.36	8.80	7.95	0.44	3
0.400	0.0400	5.43	7.19	6.15	0.41	4
0.600	0.0400	4.01	5.10	4.24	0.29	3
0.800	0.0400	1.63	2.59	2.02	0.23	4

Considering the data in Table 2.4, the number of measurements for each of the reported mean values is indicated in the last column, and the correspondent standard deviation is presented in the sixth column. The diffusion coefficient of pure (>99%) glycerol in water has been studied by many groups. The reported value at 25 °C ranges between $0.93 \times 10^{-5} \, \text{cm}^2/\text{s}$ [59] and $1.02 \times 10^{-5} \, \text{cm}^2/\text{s}$ [60].

Equations (2.3)–(2.6) and (2.8) were presented for diffusion through a homogeneous tissue slab. In reality, due to its fibrous structure, a biological tissue can be described as a porous material, and the diffusion coefficient of a chemical agent must be corrected to account for the porosity [10]:

$$D_a = \frac{D_{ai}}{p}, \tag{2.9}$$

where D_{ai} is the chemical agent diffusion coefficient within the interstitial fluid and p is the porosity coefficient, defined as:

$$p = \frac{V - V_C}{V}, \tag{2.10}$$

where V is the volume of the sample and V_C is the volume of the collagen fibers.

Figure 2.2 illustrates relations of tissue porosity p with the diffusion time τ of an OCA, which significantly decreases with a stronger porosity, and tissue sample transparency with light scattering, which makes tissue transparency much lower for strong scattering [61].

If the agent has bigger molecules, to describe their diffusion, the theory of hindered diffusion through a permeable membrane needs to be used [45–48]. From Fick's law (Eq. 2.2), the flux of matter, J (mol cm^2/s), is related with the gradient of its concentration through the diffusion coefficient [10]:

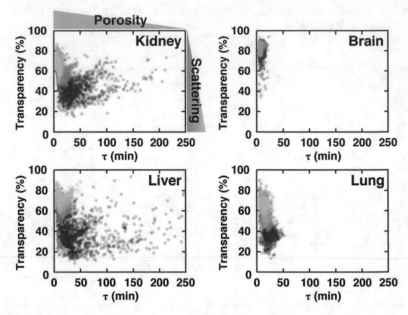

Fig. 2.2 Scatterplot of τ-transparency from samples (1 mm in thickness) of four types of tissues; for RI matching to achieve clearer images, CUBIC mount solution [250 g sucrose (50%, w/v), 125 g urea (25%, w/v), and 125 g N,N,N',N'-tetrakis(2-hydroxypropyl)ethylenediamine (25%, w/v) dissolved in 150 ml of dH$_2$O and brought up to 500 ml] was used; prior RI matching electrophoretic tissue clearing (ETC) in sodium dodecyl sulfate (SDS) solution was provided during 0, 1, 3, 6, 12, or 24 h to remove lipid from the tissue. Association of tissue porosity (induced by application of SDS via ETC protocol) with OCA diffusion time τ and tissue optical transparency (like T_c) with light scattering is shown on the upper right panel. (Reprinted from Ref. [61] under Creative Commons Attribution 4.0 International License)

$$J = -D_a \frac{\partial C}{\partial x}. \qquad (2.11)$$

Considering a thin membrane of thickness l, the stationary transport of matter from one side to the other is described by:

$$J = P_a(C_1 - C_2), \qquad (2.12)$$

where $P_a = D_a/l$ is the permeability coefficient and C_1 and C_2 are the concentrations of molecules on both sides of the membrane.

Rearranging Eqs. (2.11) and (2.12) into Eq. (2.13), the variations in the concentration of molecules in a closed space with volume V, surrounded by a permeable membrane with superficial area S, can be determined [48]:

$$\frac{\partial C_2}{\partial t} = \frac{P_a S}{V}(C_1 - C_2). \tag{2.13}$$

When the immersion solution has a large volume and C_1 can be considered as a constant, Eq. (2.13) has an approximate exponential solution, similar to Eq. (2.6). In this case, $C_2 = C_a$, $C_1 = C_a$, and

$$\tau = \frac{V}{P_a S} \text{ or } \frac{Rl}{3D_a} \text{ for a spherical membrane} \tag{2.14}$$

with R representing the radius of the membrane (of a cell or tumor necrotic core).

During chemical agent diffusion into the tissue, the RI of the background (interfibrillar) media, n_0, is a time-dependent function of the agent concentration that increases during treatment, $C_a(t)$, and is defined by Eq. (2.6). The volume fraction of the agent within the tissue sample f_a is also time-dependent and proportional to its concentration, C_a. Consequently, using Gladstone and Dale law of mixtures [10, 62–65], we can determine the RI of the background material in the tissue as:

$$n_0(t) = n_{0i}(t)f_0(t) + n_a f_a(t). \tag{2.15}$$

where $f_0(t) + f_a(t) = 1$, n_{0i} is the initial (intrinsic) RI of the ground medium in natural state. When non-osmotic or low-osmotic agents are used to treat a tissue, the intrinsic RI of the interfibrillar space can be considered as not time-dependent: $n_{0i}(t) \approx n_{0i}(t = 0)$.

For a system of noninteracting thin cylinders with a number of fibrils per unit area, ρ_s, the expression for the scattering coefficient has the form [32, 46, 47]:

$$\mu_s \cong \rho_s \left(\frac{\pi^5 a^4 n_0^3}{\lambda_0^3}\right) (m^2 - 1)^2 \left[1 + \frac{2}{(m^2 + 1)^2}\right], \tag{2.16}$$

where $\rho_s = f_{cyl}/\pi a^2$, f_{cyl} is the surface fraction of the cylinders' areas, a is the cylinder radius, $m = n_s/n_0$ (Eq. 1.17) is the relative index of refraction of cylinders (scatterers) to the background (interfibrillar space), and λ_0 is the wavelength in vacuum.

If, instead of cylinders, the tissue is composed from spherical particles having equal diameter, $2a$, a similar equation to Eq. (2.16) can be presented as the correspondent to Eq. (1.18) for μ_s via $(1 - g)$ [32, 66, 67]:

$$\mu_s = \frac{3.28\pi a^2 \rho_s}{(1 - g)} \left(\frac{2\pi a}{\lambda}\right)^{0.37} (m - 1)^{2.09}. \tag{2.17}$$

Returning back to case of the muscle and considering a first approximation, we can assume that chemicals entering the interfibrillar space will not change the radii of the scatterers (fibrils) and their density—tissue swelling or shrinkage does not take place. Consequently, the absolute change in n_0 is not very high, and variations in μ_s are caused only by the change in the relative RI of the interstitial space with respect to the RI of the scatterers—m.

Considering that a majority of tissues have m close to unity (so-called "optically soft" particles), the ratio of the scattering coefficients at a particular wavelength can be written as [46, 47]:

$$\frac{\mu_{s2}}{\mu_{s1}} \cong \left(\frac{m_2 - 1}{m_1 - 1}\right)^2.$$

(2.18)

Variations in the RI of scatterers, of the background, or both lead to a RI match or mismatch, and the correspondent changes in tissue scattering properties are quantified by this relation. According to the square dependence in Eq. (2.18), the sensitivity to index matching is very high. For instance, if $m_1 = 1.1$ and $m_2 = 1.01$, then $\mu_{s2} \cong 0.01\,\mu_{s1}$.

The RI of the scatterers, n_s, is usually kept constant during an optical clearing treatment. This means that we can use Eq. (2.15) to rewrite Eq. (2.18) in a form that is specific for tissue impregnation by an agent with a weak hyperosmolarity [32]:

$$\mu_s(t) = \mu_s(t = 0) \times \frac{\left[\frac{n_s}{n_0(t)} - 1\right]^2}{\left[\frac{n_s}{n_0(t=0)} - 1\right]^2}.$$

(2.19)

When T_c measurements are performed during the immersion treatment, Eq. (2.19) can be modified to estimate the time dependence for the RI of the ISF [28]. Usually, the immersion agents do not have strong absorption bands within the wavelength of interest, and consequently during the treatment, the absorption coefficient presents zero or very small variations when compared to the change in the scattering coefficient [28, 32]. For this case, the scattering coefficient can be calculated directly through Eq. (2.1) from T_c and thickness measurements made during the treatment [28]. If the agent used in the treatment is a hyperosmotic, it will also interact with tissue components, and tissue shrinkage or swelling might take place. For slab-form tissue samples, as presented in Fig. 2.1, the superficial area of the sample is much larger than sample thickness. Consequently, variations in sample volume can be calculated directly from sample thickness measurements made during treatment to be taken into account in Eq. (2.19) for a correct calculation. Considering such volume variations, the calculation of the time dependence for the mean RI of the ISF can be made using the following rearranged form of Eq. (2.19) [28]:

$$\bar{n}_0(t) = \frac{n_s}{\left(\sqrt{\frac{\mu_s(t) \times d(t)}{\mu_s(t=0) \times d(t=0)}} \times \left(\frac{n_s}{n_0(t=0)} - 1\right) + 1\right)}.$$

(2.20)

The introduction of the sample thickness for natural tissue, $d(t = 0)$, and during treatment $d(t)$ in Eq. (2.20) is necessary to account for scatterer density change during the treatment. If the wavelength dependence of the optical properties for the untreated tissue and experimental wavelength dependence T_c data is available for the treatment, the variations in the wavelength dependence for the RI of ISF can be calculated with Eq. (2.20). For such calculation, it is also necessary to know the wavelength dependencies for the RI of the scatterers and of the ISF in intrinsic (initial, i.e., before treatment) condition of the tissue. Such curves can be estimated if the total and mobile water contents in the tissue are known [68].

Once the time dependence for the RI of the ISF is calculated, we can use Gladstone and Dale law of mixtures (Eq. 2.15) to calculate the time dependence for the RI of the entire tissue under treatment. To perform such calculation, the time dependence for the volume fractions of scatterers and ISF is also necessary, but since the absolute volume of the scatterers is assumed to remain unchanged during treatment, both volume fractions can also be calculated from the thickness measurements over the time of treatment [28]. Such calculations allow one to quantify the magnitude of the RI matching mechanism of optical clearing.

There are several calculations that can be made with experimental data obtained during optical immersion clearing treatments, and the information produced in those calculations is very useful for a large number of ex vivo or in vivo applications. We will discuss those measurements and calculations in Chaps. 4–6. In the following chapter, we will present some typical optical clearing agents (OCAs) and their characteristics and clearing potential.

References

1. H.-U. Dodt, U. Leischner, A. Schierloh, N. Järling, C.P. Mauch, K. Deininger, J.M. Deussing, M. Eder, W. Zieglgänsberger, K. Becker, Ultramicroscopy: three dimensional visualization of neuronal networks in the whole mouse brain. Nat. Methods 4(4), 331–336 (2007)
2. A.Y. Sdobnov, M.E. Darvin, E.A. Genina, A.N. Bashkatov, J. Lademann, V.V. Tuchin, Recent progress in tissue clearing for spectroscopic application. Spectrochim. Acta A Mol. Biomol. Spectrosc. 197, 216–229 (2018)
3. A. Malpica, M. Follen, Near real time confocal microscopy of amelanotic tissue: dynamics of aceto-whitening enable nuclear segmentation. Opt. Express 6(2), 40–48 (2000)
4. R.A. Drezek, T. Collier, C.K. Brookner, A. Malpica, R. Lotan, R.R. Richards-Kortum, M. Follen, Laser scanning confocal microscopy of cervical tissue before and after application of acetic acid. Am. J. Obstet. Gynecol. 182(5), 1135–1139 (2000)
5. B. Pogue, H.B. Kaufman, A. Zelenchuk, W. Harper, G.C. Burke, E.E. Burke, D.M. Harper, Analysis of acetic-induced whitening of high-grade squamous intraepithelial lesions. J. Biomed. Opt. 6(4), 397–403 (2001)
6. C.J. Balas, G.C. Themelis, E.P. Prokopakis, I. Orfanudaki, E. Koumantakis, E. Helidonis, In vivo detection and staging of epithelial dysplasias and malignancies based on the quantitative assessment of acetic acid-tissue interaction kinetics. J. Photochem. Photobiol. B Biol. 53(1–3), 153–157 (1999)
7. G. Longcroft-Wheaton, P. Bhandari, Acetowhitening as a novel diagnostic tool for the diagnosis and characterisation of neoplasia within Barrett's oesophagus. Gut 61, A258 (2012)

8. K. Gutiérrez-Fragoso, H.G. Acosta-Mesa, N. Cruz-Ramírez, R. Hernández-Jiménez, Automatic classification of acetowhite temporal patterns to identify precursor lesions of cervical cancer. J. Phys. Conf. Ser. **475**(1), 012004-1–012004-10 (2013)
9. T.T. Wu, J.Y. Qu, Assessment of the relative contribution of cellular components to the acetowhitening effect in cell cultures and suspensions using elastic light-scattering spectroscopy. Appl. Opt. **46**(21), 4834–4842 (2007)
10. V.V. Tuchin, *Optical Clearing of Tissues and Blood* (SPIE Press, Bellingham, 2006)
11. J. Lin, S. The, W. Zheng, Z. Huang, Multimodal nonlinear optical microscopic imaging provides new insights into acetowhitening mechanisms in live mammalian cells without labeling. Biomed. Opt. Express **5**(9), 3116–3122 (2014)
12. O. Marina, A. Trujillo, C. Sanders, K. Burnett, J.P. Freyer, J.R. Mourant, The effect of acetic acid on mammalian cells, in *Biomedical Optics and 3-D Imaging*, OSA Technical Digest (CD), (Optical Society of America, Washington, DC, 2010), p. BSuD74
13. R. Drezek, A. Dunn, R. Richards-Kortum, Light scattering from cells: finite-difference time-domain simulations and goniometric measurements. Appl. Opt. **38**(16), 3651–3661 (1999)
14. A.J. Welch, M.J.C. van Gemert, *Optical-Thermal Response of Laser-Irradiated Tissue*, 2nd edn. (Springer, Dordrecht, 2011)
15. E.D. Jansen, T.G. van Leeuwen, M. Motamedi, C. Borst, A.J. Welch, Temperature dependence of the absorption coefficient of water for midinfrared laser radiation. Laser Surg. Med. **14**(3), 258–268 (1994)
16. B.I. Lange, T. Brendel, G. Hüttmann, Temperature dependence of light absorption in water at holmium and thulium laser wavelengths. Appl. Opt. **41**(27), 5797–5803 (2002)
17. L. Cordone, A. Cupane, M. Leone, E. Vitrano, Optical absorption spectra of deoxy- and oxyhemoglobin in the temperature range 300–320 K. Biophys. Chem. **24**(3), 259–275 (1986)
18. P.L. San Biagio, E. Vitrano, A. Cupane, F. Madonia, M.U. Palma, Temperature induced difference spectra of oxy- and deoxy-hemoglobin in the near IR, visible and Soret regions. Biochem. Biophys. Res. Commun. **77**(4), 1158–1165 (1977)
19. J.M. Steinke, A.P. Sheperd, Effects of temperature on optical absorbance spectra of oxy-, carboxy- and deoxy-hemoglobin. Clin. Chem. **38**(7), 1360–1364 (1992)
20. R. Sfareni, A. Boffi, V. Quaresima, M. Ferrari, Near infrared absorption spectra of human deoxy- and oxyhemoglobin in the temperature range 20–40 degrees C. Biochim. Biophys. Acta **1340**(2), 165–169 (1997)
21. K. Gray, E.F. Slade, The temperature dependence of the optical absorption spectra of some methaemoglobin derivatives. Biochem. Biophys. Res. Commun. **48**(4), 1019–1024 (1972)
22. J.F. Black, N. Wade, J.K. Barton, Mechanistic comparison of blood undergoing laser photocoagulation at 532 and 1064 nm. Lasers Surg. Med. **36**(2), 155–165 (2005)
23. J.F. Black, J.K. Barton, Chemical and structural changes in blood undergoing laser photocoagulation. Photochem. Photobiol. **80**(1), 89–97 (2004)
24. A. Kienle, R.A. Hibst, New optimal wavelength for treatment of port wine stains? Phys. Med. Biol. **40**(10), 1559–1576 (1995)
25. W. Jia, B. Choi, W. Franco, J. Lotfi, B. Majaron, G. Aguilar, J.S. Nelson, Treatment of cutaneous vascular lesions using multiple-intermittent cryogen spurts and two-wavelength laser pulses: numerical and animal studies. Laser Surg. Med. **39**(6), 494–503 (2007)
26. I.F. Cilesiz, A.J. Welch, Light dosimetry: effects of dehydration and thermal damage on the optical properties of the human aorta. Appl. Opt. **32**(4), 477–487 (1993)
27. S. Jaywant, B. Wilson, M. Patterson, L. Lilge, T. Flotte, J. Woolsey, C. McCulloch, Temperature dependent changes in the optical absorption and scattering spectra of tissues: correlation with ultrastructure. Proc. SPIE **1882**, 218–229 (1993)
28. L. Oliveira, M.I. Carvalho, E. Nogueira, V.V. Tuchin, Skeletal muscle dispersion (400–1000 nm) and kinetics at optical clearing. J. Biophotonics **11**(1), e201700094 (2018)
29. H.S. Dhadwal, R.R. Ansari, M.A. DellaVecchia, Coherent fiber optic sensor for early detection of caractogenesis in the human eye lens. Opt. Eng. **32**(2), 233–238 (1993)

30. B. Grzegorzewski, S. Yermolenko, Speckle in far-field produced by fluctuations associated with phase separation. Proc. SPIE **2647**, 343–349 (1995)
31. B. Choi, T.E. Milner, J. Kim, J.N. Goodman, G. Vargas, G. Ahuilar, J.S. Nelson, Use of optical coherence tomography to monitor biological tissue during cryosurgery. J. Biomed. Opt. **9**(2), 282–286 (2004)
32. V.V. Tuchin, *Tissue Optics – Light Scattering Methods and Instruments for Medical Diagnosis*, 3rd edn. (SPIE Press, Bellingham, 2015)
33. P.O. Rol, *Optics for Transscleral Laser Applications*, Dissertation for the degree of Doctor of Natural Sciences, N9655, Swiss Federal Institute of Technology, Zurich (1992), p. 152
34. E.K. Chan, B. Sorg, D. Protsenko, M. O'Neil, M. Motamedi, A.J. Welch, Effects of compression on soft tissue optical properties. IEEE J. Select. Top. Quant. Electron. **2**(4), 943–950 (1996)
35. A.A. Gurjarpadhye, W.C. Vogt, Y. Liu, C.G. Rylander, Effect of localized mechanical indentation on skin water evaluated using OCT. Int. J. Biomed. Imag. **2011**, 817250-1–817250-8 (2011)
36. M.Y. Kirillin, P.D. Agrba, V.A. Kamensky, In vivo study of the effect of mechanical compression on formation of OCT images of human skin. J. Biophotonics **3**(12), 752–758 (2010)
37. W. Spalteholz, *Über das Durchsichtigmachen von menschlichen und tierichen Präparaten und seine theoretischen Bedingungen, nebst Anhang: Über Knochenfärbung* (S. Hirzel, Leipzig, 1911)
38. W. Spalteholz, *Über das Durchsichtigmachen von menschlichen und tierichen Präparaten und seine theoretischen Bedingungen, nebst Anhang: Über Knochenfärbung* (S. Hirzel, Leipzig, 1914)
39. D.S. Richardson, J.W. Lichtman, Clarifying tissue clearing. Cell **162**(2), 246–257 (2015)
40. A. Azaripour, T. Lagerweij, C. Scharfbillig, A.E. Jadczak, B. Willershausen, C.J.F. van Noorden, A survey of clearing techniques for 3D imaging of tissues with special reference to connective tissue. Prog. Histochem. Cytochem. **51**, 9–23 (2016)
41. E.A. Genina, A.N. Bashkatov, Y.P. Sinichkin, I.Y. Yanina, V.V. Tuchin, Optical clearing of biological tissues: prospects of application in medical diagnosis and phototherapy. J. Biomed. Photon. Eng. **1**(1), 22–58 (2015)
42. D. Zhu, K.V. Larin, Q. Luo, V.V. Tuchin, Recent progress in tissue optical clearing. Laser Photon. Rev. **7**(5), 732–757 (2013)
43. M.S.C. Kauhanen, A.M. Salmi, E.K. Von Boguslawsky, I.V.V. Leivo, S.L. Asko-Seljavaara, Muscle fiber diameter and muscle type distribution following free microvascular muscle transfers: a prospective study. Microsurgery **18**(2), 137–144 (1998)
44. W.L. Bragg, A.B. Pippard, The form birefringence of macromolecules. Acta Cryst. **6**, 865–867 (1953)
45. L. Oliveira, A. Lage, M. Pais Clemente, V.V. Tuchin, Optical characterization and composition of abdominal wall muscle from rat. Opt. Laser. Eng. **47**(6), 667–672 (2009)
46. V.V. Tuchin, I.L. Maksimova, D.A. Zimnyakov, I.L. Kon, A.H. Mavlutov, A.A. Mishin, Light propagation in tissues with controlled optical properties. J. Biomed. Opt. **2**(4), 401–417 (1997)
47. V. Tuchin, I. Maksimova, D. Zimnyakov, I. Kon, A. Mavlutov, A. Mishin, Light propagation in tissues with controlled optical properties. Proc. SPIE **2925**, 118–132 (1996)
48. A. Kotyk, K. Janacek, *Membrane Transport: An Interdisciplinary Approach* (Plenum Press, New York, 1997)
49. A.N. Bashkatov, E.A. Genina, V.V. Tuchin, Measurement of glucose diffusion coefficients in human tissues, Chapter 19, in *Handbook of Optical Sensing of Glucose in Biological Fluids and Tissues*, ed. by V. V. Tuchin, (Taylor & Francis Group LLC, CRC Press, London, 2009), pp. 87–621
50. A.N. Bashkatov, E.A. Genina, Y.P. Sinichkin, V.I. Kochubey, N.A. Lakodina, V.V. Tuchin, Glucose and mannitol diffusion in human dura mater. Biophys. J. **85**(5), 3310–3318 (2003)
51. L.M. Oliveira, M.I. Carvalho, E.M. Nogueira, V.V. Tuchin, The characteristic time of glucose diffusion measured for muscle tissue at optical clearing. Laser Phys. **23**(7), 075606-1–075606-6 (2013)

52. L.M. Oliveira, M.I. Carvalho, E.M. Nogueira, V.V. Tuchin, Diffusion characteristics of ethylene glycol in skeletal muscle. J. Biomed. Opt. **20**(5), 051019-1–051019-10 (2015)
53. L.M. Oliveira, M.I. Carvalho, E.M. Nogueira, V.V. Tuchin, Optical clearing mechanisms characterization in muscle. J. Innov. Opt. Health Sci. **9**(5), 1650035-1–1650035-19 (2016)
54. S. Carvalho, N. Gueiral, E. Nogueira, R. Henrique, L.M. Oliveira, V.V. Tuchin, Glucose diffusion in colorectal mucosa – a comparative study between normal and cancer tissues. J. Biomed. Opt. **22**(9), 091506-1–091506-12 (2017)
55. D.J. Tomlinson, Temperature dependent self-diffusion coefficient measurements of glycerol by pulsed N.M.R. technique. Mol. Phys. **25**(3), 735–738 (1972)
56. F. Mallamace, C. Corsaro, D. Mallamace, E. Vasi, C. Vasi, H.E. Stanley, Some considerations on the transport properties of water-glycerol suspensions. J. Chem. Phys. **144**, 014501 (2016)
57. M.A. Araújo, E.C. Ferreira, A.M. Cunha, M. Mota, Determination of diffusion coefficients of glycerol and glucose from starch based thermoplastic compounds on stimulated physiological solution. J. Mater. Sci. Mater. Med. **16**(3), 239–246 (2005)
58. G. Ternström, A. Sjöstrand, G. Aly, Å. Jernqvist, Mutual diffusion coefficients of water + ethylene glycol and water + glycerol mixtures. J. Chem. Eng. Data **41**(4), 876–879 (1996)
59. A.L. Weber, Kinetics of organic transformations under mild aqueous conditions: implications for the origin of life and its metabolism. Orig. Life Evol. Biosph. **34**(5), 473–495 (2004)
60. G. D'Errico, O. Ortona, F. Capuano, V. Vitagliano, Diffusion coefficients for the binary system glycerol + water at 25°C. A velocity correlation study. J. Chem. Data **49**, 1665–1670 (2004)
61. J.H. Kim, M.J. Jang, J. Choi, E. Lee, K.D. Song, J. Cho, K.T. Kim, H.J. Cha, W. Sun, Optimizing tissue-clearing conditions based on analysis of the critical factors affecting tissue clearing procedures. Sci. Rep. **8**(1), 12815 (2018)
62. D.W. Leonard, K.M. Meek, Refractive indices of the collagen fibrils and extrafibrillar material of the corneal stroma. Biophys. J. **72**(3), 1382–1387 (1997)
63. K.M. Meek, S. Dennis, S. Khan, Changes in the refractive index of the stroma and its extrafibrillar matrix when the cornea swells. Biophys. J. **85**(4), 2205–2212 (2003)
64. K.M. Meek, D.W. Leonard, C.J. Connon, S. Dennis, S. Khan, Transparency, swelling and scarring in the corneal stroma. Eye **17**(8), 927–936 (2003)
65. O. Zhernovaya, O. Sydoruk, V. Tuchin, A. Douplik, The refractive index of human hemoglobin in the visible range. Phys. Med. Biol. **56**(13), 4013–4021 (2011)
66. R. Graaff, J.G. Aarnoudse, J.R. Zijp, P.M.A. Sloot, F.F.M. De Mul, J. Greve, M.H. Koelink, Reduced light-scattering properties for mixtures of spherical particles: a simple approximation derived from Mie calculations. Appl. Opt. **31**(10), 1370–1376 (1992)
67. H. Liu, B. Beauvoit, M. Kimura, B. Chance, Dependence of tissue optical properties on solute-induced changes in refractive index and osmolarity. J. Biomed. Opt. **1**(2), 200–211 (1996)
68. I. Carneiro, S. Carvalho, R. Henrique, L. Oliveira, V.V. Tuchin, Water content and scatterers dispersion evaluation in colorectal tissues. J. Biomed. Photon. Eng. **3**(4), 040301-1–040301-10 (2017)

Chapter 3
Typical Optical Clearing Agents

3.1 Desirable Properties of Optical Clearing Agents

According to the explanation given in Chap. 1, the high scattering properties observed in biological tissues are in great part created by the refractive index (RI) mismatch between the various tissue components and fluids. The large water content in tissue fluids is the main responsible for this effect, since the RI of water is low, when compared to the RI values observed for other tissue components, such as protein fibers, lipid globules, cell organelles, etc. [1, 2].

By replacing the water in the tissue fluids by a harmless liquid that has a higher RI, better matched to the one of the other tissue components, it is possible to reduce the RI mismatch and turn the tissues to be clear. The research related to optical clearing (OC) has grown fast in the past two decades [3–5], where studies have been conducted to evaluate the clearing potential of different agents in various tissues. Optical clearing agents (OCAs) for in vivo applications are harmless chemicals, and the most common are grouped in three classes: sugars, alcohols, and electrolyte contrasting solutions [1, 4, 6, 7]. Table 3.1 presents some typical OCAs within these classes and related data at 20 °C.

The sugars presented in Table 3.1 are found in solid form as powder after refined, but they are soluble in water and other liquids. They have various applications in food industry as sweeteners and can be used also in medicine applications.

The alcohols in Table 3.1 (sorbitol, glycerol, and PG) are used in food industry, cosmetics, or medicine [1] and are very valuable in OC treatments due to their high RIs and solubility in water, allowing to prepare solutions with high osmolarity.

The Trazograph^{TM} is a derivative of 2,4,6-triiodobenzene acid. It is a water-soluble colorless liquid that is produced in concentrations of 60% or 76%. It is used as an intravenous radiopaque (X-ray contrast) agent, but since it has high RI and allows one to prepare solutions with high osmolarity, it is also valuable for OC treatments [1]. Similar to Trazograph^{TM}, Hypaque^{TM} and Verografin^{TM} are also X-ray contrast agents, which are also valuable for OC treatments [1].

L. M. C. Oliveira, V. V. Tuchin, *The Optical Clearing Method*,
SpringerBriefs in Physics, https://doi.org/10.1007/978-3-030-33055-2_3

Table 3.1 Typical OCAs (data at 20 °C)

	Agent	Chemical formula	Molecular weight (g/mol)	RI ($\lambda = 589$ nm)	Concentration (in water) (%)	Reference
Sugars	Glucose	$C_6H_{12}O_6$	180.2	1.4148	54	[8]
	Sucrose	$C_{12}H_{22}O_{11}$	342.3	1.4906	80	
	Fructose	$C_6H_{12}O_6$	180.2	1.4141	48	
Alcohols	Sorbitol	$C_6H_{14}O_6$	182.2	1.375	25	[1]
	Glycerol	$C_3H_8O_3$	92.1	1.4747	100	
	Propylene glycol (PG)	$C_3H_8O_2$	76.1	1.4326	100	
Electrolyte solutions	Trazograph™-76[a]	$C_{11}H_8I_3N_2NaO_4/$ $C_{18}H_{26}I_3N_3O_9$	635.89/809.13	1.460	100	
	Hypaque™-60[a]	$C_{11}H_8I_3N_2NaO_4/$ $C_{18}H_{26}I_3N_3O_9$	635.89/809.13	1.437	100	
	Verografin™-76[a]	$C_{11}H_8I_3N_2NaO_4/$ $C_{18}H_{26}I_3N_3O_9$	635.89/809.13	1.485	100	

[a]Radiopaque (X-ray contrast) substances Trazograph™, Hypaque™, Verografin™, Urografin™, Urostras™, and Uropolinum™ are synonyms; the release form is of 60% and 76% solution of mixture of diatrizoate sodium (635.89 g/mol) and/or diatrizoate meglumine (809.13 g/mol)

The Omnipaque™ (iohexol) pertains to a new generation of nonionic, water-soluble radiographic contrast media. In aqueous solution each triiodinated molecule remains undissociated; thus osmolalities comparable with blood plasma (285 mOsm/kg water) and cerebral spinal fluid (CSF) (301 mOsm/kg water) can be provided (see Table 3.2). Therefore intrathecal, intravascular, and oral/body cavity use are possible with no complications. The chemical formula of iohexol is $C_{19}H_{26}I_3N_3O_9$; its molecular weight is 821.14 with iodine content of 46.36% [9]. For OC, agent refractivity is also important, for Omnipaque™ (300) with concentration of iodine of 300 mgI/ml, RI is 1.432 ($\lambda = 785$ nm) [10].

In addition to the OCAs presented in Tables 3.1 and 3.2, there are many others that have been tested and produced significant clearing effects in various biological materials. Several of those OCAs, including natural substances, carbohydrates, diuretics, oils, serums, and other products, are presented in Table 1 of Ref. [1].

In some treatments and experimental studies, it is preferable to use diluted solutions of these OCAs [11–13]. A method to prepare these solutions will be described in Sect. 3.4.

For an OC treatment to turn a tissue clear, it is desirable that the OCA has a high transmittance and no significant absorption bands in the spectral range of interest. The RI of the OCA should also approximate the one of tissue scatterers in the same spectral range. The diffusivity of the OCA into the tissue should also be high, but such property depends both on the molecular size of the OCA and on tissue permeability. We will discuss all these desirable characteristics for OCAs in the following sections, with the exception of the evaluation of OCA diffusion properties in tissues, which will be described in detail in Chap. 6.

It should be mentioned that more complex OCAs designed mostly for in vitro biological applications providing a high degree of OC for fixed tissues during prolonged clearing time (hours and days) are also available [14]. These multicomponent agents and corresponding protocols, such as 3DISCO (3D imaging of solvent-cleared organs), uDISCO (ultimate DISCO), SeeDB (see deep brain), ScaleS, Clear T2, and PACT (passive CLARITY technique), are used for 3D imaging of solvent-cleared different small animal organs including deep brain optical imaging. More than 1600 chemicals were screened recently by a high-throughput evaluation system for each chemical process to present a complete chemical landscape for tissue OC based on hydrophilic reagents using CUBIC (clear, unobstructed brain/body imaging cocktails and computational analysis) protocols [15].

Table 3.2 Physical properties of Omnipaque™ (iohexol) [9]

Concentration (mgI/ml)	Osmolality (mOsm/kg water)	Osmolarity (mOsm/l)	Absolute viscosity (cp)		Specific gravity 37 °C
			20 °C	37 °C	
140	322	273	2.3	1.5	1.164
180	408	331	3.1	2.0	1.209
240	520	391	5.8	3.4	1.280
300	672	465	11.8	6.3	1.349
350	844	541	20.4	10.4	1.406

3.2   Absorption and Dispersion Data for OCAs

OCAs are in general translucent liquids, but they have some absorption bands. The RI of these agents is known to decrease with increasing wavelength. When performing tissue spectroscopy, it is important that the replacement of tissue water by an OCA does not modify the spectral form of the tissue. To prevent this, the absorption bands of OCAs should not be located in spectral ranges where the RI of the tissue is low. We have gathered some data that allows us to make this analysis for some OCAs.

Since some OCAs are used in other applications and in other fields of research, their dispersion data is of high interest. Consequently, such data has been acquired for some OCAs at room temperature (20–25 °C) and made available online [16]. Such data can be used to reconstruct the dispersion curves for OCAs in some wavelength ranges. Figure 3.1a contains the wavelength dependence for the RI of dimethyl sulfoxide (DMSO) [17], propylene glycol (PG) [18], glycerol [19], and ethylene glycol (EG) [20] for the wavelength range between 200 and 1000 nm. Figure 3.1b contains RI data for glycerol; PEG-300 (polyethylene glycol); lactulose; electronic cigarette fluid (e-cig), which is composed of 40% PG and 60% glycerol; Omnipaque-300 (radiology contrast agent); PG; 40%-glucose; and 60%-glucose (aqueous solutions containing different percentages of glucose—40% and 60%). These data were measured for wavelengths between 480 and 1550 nm. PEG-300 and lactulose were measured at 21 °C, glycerol was measured at 23 °C, e-Cyg was measured at 24 °C, and the remaining OCAs were measured at 27 °C.

The datasets or curves for each OCA presented in Fig. 3.1 concern different wavelength ranges and different temperatures, but we see the decreasing tendency with increasing wavelength in all cases. The empirical relations that best fit such dispersions for various materials and fluids, including biological, are the Cauchy, Hartmann, Conrady, Sellmeier, and some other formulas [21–23], such as the Cornu relation [7, 24, 25]. For all the datasets presented in Fig. 3.1a, the best fitting was obtained with the Cornu relation.

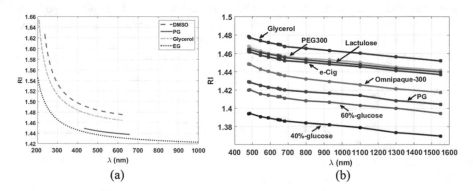

Fig. 3.1 Wavelength dependence for the RI of some OCAs: data from Ref. [16] (20–25 °C) (**a**) and experimental data measured by our colleague Dr. Ekaterina Lazareva at Saratov State University using commercial multiwavelength refractometer Atago DR-M2/1550 (**b**)

The data for DMSO presented in Fig. 3.1a corresponds to a temperature of 22 °C, and it is available only for a wavelength range between 236 and 641 nm [17]. Using that experimental data, we calculated the dispersion curve for DMSO (R-square $= 0.9996$) presented in Fig. 3.1.

$$n_{\text{DMSO}}(\lambda, 22°C) = 1.454 + \frac{9.521}{\lambda - 180.7}. \tag{3.1}$$

For the case of PG in Fig. 3.1a, data is available only for four wavelengths between 434 and 656 nm and for a temperature of 25 °C [18]. The curve for PG presented in Fig. 3.1a corresponds to the fitting of this experimental data (R-square $= 0.9998$), as described by Eq. (3.2).

$$n_{\text{PG}}(\lambda, 25°C) = 1.4234 + \frac{7.214}{\lambda - 152.7}. \tag{3.2}$$

In the case of glycerol, the experimental data is available at room temperature (20 °C) and between 206 and 620 nm [19]. The curve presented in Fig. 3.1a for glycerol was obtained by fitting that data (R-square $= 0.9937$) with Eq. (3.3).

$$n_{\text{Glycerol}}(\lambda, 20°C) = 1.444 + \frac{9.534}{\lambda - 160.5}. \tag{3.3}$$

Experimental data for EG is available for a very wide wavelength range (~185 nm–55 μm), also at room temperature (20 °C) [20], from which we retrieved data only between 200 and 1000 nm. The curve presented in Fig. 3.1a corresponds to the fitting of that experimental data (R-square $= 0.9993$) with Eq. (3.4).

$$n_{\text{EG}}(\lambda, 20°C) = 1.413 + \frac{8.763}{\lambda - 133.5}. \tag{3.4}$$

Similar fittings with Cornu-type equation were made for the experimental data presented in Fig. 3.1b, obtaining also good R-square values in all cases. Equation (3.5) was obtained to fit the experimental RI data of glycerol (R-square $= 0.9886$).

$$n_{\text{Glycerol}}(\lambda, 23°C) = 1.434 + \frac{35.78}{\lambda + 352.9}. \tag{3.5}$$

For PEG-300, we fitted the experimental RI data with Eq. (3.6) (R-square $= 0.9772$).

$$n_{\text{PEG-300}}(\lambda, 21°C) = 1.428 + \frac{26.94}{\lambda + 222}. \tag{3.6}$$

In the case of lactulose, the RI data fitting is described by Eq. (3.7) (R-square $= 0.9933$).

$$n_{\text{lactulose}}(\lambda, 21°C) = 1.418 + \frac{46.07}{\lambda + 530.9}. \tag{3.7}$$

For the case of e-cig, the data fitting was obtained with Eq. (3.8) (R-square $= 0.9849$).

$$n_{\text{e-Cig}}(\lambda, 24°C) = 1.418 + \frac{40.5}{\lambda + 460}. \tag{3.8}$$

The experimental data for Omnipaque-300 was fitted with Eq. (3.9) (R-square $= 0.9939$).

$$n_{\text{Omnipaque-300}}(\lambda, 27°C) = 1.402 + \frac{27.44}{\lambda + 119.1}. \tag{3.9}$$

Equation (3.10) was obtained while fitting the experimental RI data of PG (R-square $= 0.9784$).

$$n_{\text{PG}}(\lambda, 27°C) = 1.378 + \frac{65.77}{\lambda + 836.4}. \tag{3.10}$$

The calculated fitting for the experimental RI data of 60%-glucose is described by Eq. (3.11) (R-square $= 0.9845$).

$$n_{60\%\text{-glucose}}(\lambda, 27°C) = 1.373 + \frac{48.36}{\lambda + 571.5}. \tag{3.11}$$

For the experimental RI data of 40%-glucose, we obtained the fitting described by Eq. (3.12) (R-square $= 0.9871$).

$$n_{40\%\text{-glucose}}(\lambda, 27°C) = 1.330 + \frac{114.9}{\lambda + 1333}. \tag{3.12}$$

In all the previous 12 equations, the wavelength λ is to be used in nanometers. As we can see from data in Fig. 3.1, the RI of OCAs decreases with wavelength for any temperature and between 200 and 1000 nm is in general well fitted with a Cornu-type equation. Looking into the experimental data presented in Fig. 3.1b, we can infer that if measurements could be performed at lower wavelengths (below 400 nm), the RI would increase abruptly, as presented in Fig. 3.1a. The same can be suggested for PG, where data is available for a reduced wavelength range (see Fig. 3.1a). All OCAs presented in Fig. 3.1 have dispersion curves above the one observed for water (see Fig. 1.5), meaning that treatments of tissues with these agents will originate an effective RI matching between tissue components and fluids.

In face of the characteristic dispersion curves presented in Fig. 3.1, we see that for these OCAs to be adequate to perform treatments of biological tissues, their absorption bands should be located in the deep ultraviolet (UV), for wavelengths below 300–350 nm, where the RI is significantly high. To identify the wavelengths where the absorption bands occur for typical agents, we have measured the collimated transmittance (T_c) spectra for a few OCAs between 200 and 1000 nm. Figure 3.2 presents those spectra for glycerol, EG, and PG as presented in https://refractiveindex.info/ and as acquired at our lab in Porto from a 2 mm thickness sample of 54%-glucose, pure glycerol, pure PG, and pure EG.

From Fig. 3.2 we see that the major absorption bands for these OCAs are located in the deep UV, where the RI is high, or in the infrared. Data in both graphs of Fig. 3.2 represent the internal T_c spectra, meaning that reference spectrum in these measurements was acquired with the entire setup, but without the OCA in the cuvette. In the graph presented in Fig. 3.2a, we see that glycerol has 0% transmittance between 100 and ~140 nm, rising strongly to 100% between 140 and 155 nm. Such transmittance keeps unchanged for wavelengths between 155 and 620 nm, which is the upper limit for the band where the data was acquired [19]. Comparing the T_c spectra for glycerol between data from Fig. 3.2a, b, we see from our measurements that glycerol shows a secondary absorption peak near 275 nm and that the major absorption band extends to higher wavelengths than in graph of Fig. 3.2a. This means that commercial glycerol we used has some impurities. From our experimental measurements, represented in Fig. 3.2b, we see that all OCAs studies show small absorption band near 980 nm, which indicates the presence of some water.

For the cases of EG and PG, we see from Fig. 3.2a that they also present a strong absorption band near 1500 nm [18, 20]. This means that at infrared wavelengths, due to OCA absorption and low RI, the RI matching mechanism must be studied carefully to evaluate its effectiveness. Considering the UV, the strong absorption bands presented in Fig. 3.2a for these two agents have high magnitude, but since

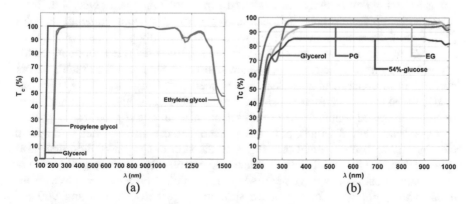

Fig. 3.2 T_c spectra of some OCAs: data from https://refractiveindex.info/ (**a**) and measured at our lab in Porto (**b**)

evidence of strongly increasing RI in this range is presented in Fig. 3.1a, we can expect a robust RI matching in tissues treated with these agents, glycerol included. Comparing between data in both graphs of Fig. 3.2 for EG and PG, we see that our experimental data matches perfectly the data in literature for PG. For the case of EG, we see that the UV band extends to nearly 400 nm for our measurements, while in the case of literature data (Fig. 3.2a), that band extends only to 300 nm. This means that our EG might also contain some impurities.

By confronting literature and experimental data in Fig. 3.2, we see that some OCAs available in the market are not entirely pure. Simple spectral T_c measurements indicate the presence of some impurities, which can be detected by their spectral signatures. Independently of OCAs containing some impurities and presenting strong absorption bands in the deep UV, the fact that the RI is strongly increasing in that range indicates that the RI matching mechanism will be maximum in tissues treated with these OCAs.

3.3   OCA Enhanced Delivery

In some cases where tissue permeability to certain agents is low, the use of OC enhancers in the treating solution becomes necessary to improve the diffusion of the agent that will perform the RI matching mechanism.

Tissue permeability to agents can be increased using various methods. Y. Damestani et al. used a combination of microneedle poration with OCA heating and pneumatic pressure application in ex vivo porcine skin samples [26]. In this study it was observed a 68% increase in transmittance of the skin for an optimized enhancement protocol of heated PG (at 45 °C) + microneedle pre-treatment +35 kPa vacuum pre-treatment +103 kPa positive pressure posttreatment. These optimal enhancement conditions also show an increase in tissue transmittance of 46% relative to PG topical application to the skin at room temperature, without any enhancement. Another study has evaluated skin dehydration and glycerol delivery through the epidermis both in intact and perforated stratum corneum samples [27]. Part of the samples in this study were photothermally perforated with a flash lamp, and it was observed that such procedure improves the water flux out and the glycerol flux through the stratum corneum skin layer. Skin or other tissue's barrier perforation to enhance optical clearing has been tested in different ways. In one study, the skin was perforated using a needle roller [28], while combination of microneedling and sonophoresis of the skin has also produced good results [29].

Laser irradiation of tissues, prior to OC treatment, can also enhance skin optical clearing. It has been reported that exposing the skin to CO_2 laser light decreases tissue reflectance for wavelengths between 400 and 600 nm [30], easing the work of consecutive treatment with glycerol solutions, which decreases tissue reflectance even further, but now for the entire spectral range between 400 and 900 nm (wavelength band used in study of Ref. [30]). A similar study has been reported [31], where a 980 nm diode laser was used to disrupt the stratum corneum barrier. As a result of this study, tissue depth for OCT signal capturing has increased 42%.

A most common method to enhance optical clearing is to use some particular chemical agents, usually designated as chemical penetration enhancers (CPEs) in the treating solution. Some of these enhancers are DMSO [1, 32, 33], oleic acid [1, 34, 35], hyaluronic acid [36], PG, azone, or thiazone [37].

More than 100 potential CPEs for transdermal molecular delivery were tested to evaluate their efficiency and irritation potential [38]. Different CPEs perturb the skin barrier via extraction or fluidization of lipid bilayers that can be accompanied by skin irritation response, which correlates with the denaturation of stratum corneum proteins. It was shown that DMSO irritation potential is comparable with oleic and linoleic acids, which are commonly used CPEs.

Due to the hygroscopic nature of some of these enhancers, the dehydration mechanism becomes faster and more efficient. We have performed a study where skeletal muscle samples were treated with a mixture containing ethanol, glycerol, and water, in the volume proportions 1:1:2 [39]. In this study, due to the fast dehydration induced by ethanol, muscle transmittance almost doubled after 2 min for the spectral range between 400 and 800 nm.

An alternative way to overcome tissue permeability to OCAs is to inject them directly inside the tissue, as reported in literature [40].

Sometimes, OCA-water solutions with different OCA osmolarities are desired for particular OC studies. It is known that oversaturated solutions stimulate and induce the unique dehydration mechanism, while solutions containing the same water as the mobile water in the tissue under treatment induce a unique OCA flow into the tissue due to the water balance between tissue and treating solution [11–13]. The treatment of tissues with different OCA osmolarities to induce those particular fluxes has particular interest to evaluate the diffusion properties of OCAs and water in a particular tissue or to discriminate pathologies. We will describe this technique in detail in Chap. 6.

3.4 Method to Prepare Aqueous Solutions with Different OCA Osmolarities

According to what has been described in the previous section, the preparation of aqueous solutions with different OCA osmolarities is necessary for certain OC studies.

A simple and precise method to prepare those diluted solutions consists on mixing distilled water with the OCA until a desired RI is obtained. It should be stressed that the RI depends on temperature, meaning that the preparation of solutions must be done at a specific temperature, for which we know the RI of the various solutions. Due to historical and instrumental reasons, the RI is well known at 589.6 nm, the reference wavelength of the Abbe refractometer. As an example, if we want to prepare an aqueous solution that contains 40% of glycerol at 20 °C, we should first fix the temperature at that value and then mix glycerol and water until we measure 1.3841 at the Abbe refractometer. If the RI value obtained at the refractometer is too high, more water should be added to the solution and mixed well before reading the RI again. In opposition, if the RI value is low, then some glycerol should be added and mixed to increase the RI of the solution. This is a trial and error method, but it is simple, inexpensive, and effective.

There are several sets of RI values for solutions with different concentrations of various OCAs in literature or the Internet, but in general they are referred to 20 °C. If other temperatures are to be considered in the preparation of solutions, the corresponding RI values should be known. Some examples are presented in the following subsections, considering the reference wavelength of 589.6 nm and a temperature of 20 °C.

3.4.1 Glucose-Water Solutions

The RI of glucose-water solutions at 20 °C and 589.6 nm can be calculated according to equation [1]:

$$n_{\text{Glucose-w}} = n_{\text{water}} + 1.515 \times 10^{-3} \times C_{\text{Glucose}}. \tag{3.13}$$

where $n_{\text{Glucose-w}}$ represents the RI of the glucose-water solution, n_{w} is the RI of water at the reference wavelength (1.3330 at 589.6 nm), and C_{Glucose} is the glucose concentration in the solution (in percentage).

If using a different reference wavelength, the RI of water can be calculated according to [1]:

$$n_{\text{w}} = 1.31848 + \frac{6.662}{\lambda - 129.2}. \tag{3.14}$$

The wavelength (λ) in Eq. (3.14) is to be used in nanometers (nm). As an example, at 20 °C and 589.6 nm, the RI of water is calculated as 1.3330. Using this value in Eq. (3.13) and replacing the glucose concentration (C_{Glucose}) by 30 (for a 30%-glucose solution), we obtain a RI for the solution equal to 1.3784. Due to glucose solubility in water at 20 °C, these solutions can be prepared for any glucose concentration until a maximum of 54%. Much higher concentrated solutions can be prepared by increasing temperature [41]. A few values for typical glucose-water solutions are presented in Table 3.3, as calculated with Eq. (3.13).

Table 3.3 RI values for typical glucose-water solutions (at 589.6 nm and 20 °C)

	Glucose concentration in solution							
	20%	25%	30%	35%	40%	45%	50%	54%
RI	1.3633	1.3708	1.3784	1.3860	1.3936	1.4011	1.4087	1.4148

3.4.2   Glycerol-Water Solutions

The RI of glycerol-water solutions is also available in the Internet [42]. From Ref. [42], the RI of glycerol-water solutions is available for each glycerol concentration between 1% and 100% at 20 °C. Considering typical solutions, we have included some data in Table 3.4.

If we represent the values in Table 3.4 in a graph, we see the linear dependency between the RI of glycerol-water solution ($n_{\text{Glycerol-w}}$) and glycerol concentration in solution (C_{Glycerol}). Such dependency is described by equation:

$$n_{\text{Glycerol-w}} = 1.3293 + 1.4 \times 10^{-3} \times C_{\text{Glycerol}}. \tag{3.15}$$

3.4.3   EG-Water Solutions

For ethylene glycol (EG)-water solutions, we can find RI data on the website of Mettler Toledo™ [43]. Once again, that data is referred to 589.6 nm and 20 °C and can be described as [44]:

$$n_{\text{EG-w}} = 1.3326 + 1.0 \times 10^{-3} \times C_{\text{EG}}, \tag{3.16}$$

where $n_{\text{EG-w}}$ represents the RI of the EG-water solution and C_{EG} represents the EG concentration in solution (in percentage). Table 3.5 contains typical values for the RI of EG-water solutions, as calculated using Eq. (3.16).

3.4.4   PG-Water Solutions

For PG-water solutions, RI data is also available for 20 °C [45]. That data can be fitted by equation:

Table 3.4 RI values for typical glycerol-water solutions (at 589.6 nm and 20 °C)

	Glycerol concentration in solution								
	20%	25%	30%	35%	40%	45%	50%	55%	60%
RI	1.3575	1.3640	1.3707	1.3774	1.3841	1.3909	1.3981	1.4055	1.4130

Table 3.5 RI values for typical EG-water solutions (at 589.6 nm and 20 °C)

	EG concentration in solution								
	20%	25%	30%	35%	40%	45%	50%	55%	60%
RI	1.3526	1.3576	1.3626	1.3676	1.3726	1.3776	1.3826	1.3876	1.3926

Table 3.6 RI values for typical PG-water solutions (at 589.6 nm and 20 °C)

	PG concentration in solution								
	20%	25%	30%	35%	40%	45%	50%	55%	60%
RI	1.3552	1.3610	1.3667	1.3725	1.3782	1.3838	1.3892	1.3944	1.3995

Table 3.7 RI values for typical sucrose-water solutions (at 589.6 nm and 20 °C)

	Sucrose concentration in solution								
	20%	25%	30%	35%	40%	45%	50%	55%	60%
RI	1.3639	1.3724	1.3812	1.3904	1.3999	1.4099	1.4201	1.4309	1.4419

$$n_{\text{PG-w}} = 1.3329 + 1.1 \times 10^{-3} \times C_{\text{PG}}, \tag{3.17}$$

where $n_{\text{PG-w}}$ represents the RI of the PG-water solution and C_{PG} represents the PG concentration in solution (in percentage). Table 3.6 contains typical values for the RI of PG-water solutions, as calculated using Eq. (3.17).

3.4.5   Sucrose-Water Solutions

Sucrose-water solutions can also be prepared, since the RI for those solutions is also available in the Internet [46]. In opposition to the previous OCAs, the dependence between the RI of the sucrose-water ($n_{\text{sucrose-w}}$) solutions and the sucrose concentration (C_{sucrose}) in solution is not linear—it grows exponentially. Equation (3.18) describes such dependency:

$$n_{\text{sucrose-w}} = 1.334 + 1.349 \times 10^{-3} \times C_{\text{sucrose}} + 7.5 \times 10^{-6} \times C_{\text{sucrose}}^2. \tag{3.18}$$

Table 3.7 presents the RI for typical sucrose-water solutions, as calculated with Eq. (3.18).

References

1. V.V. Tuchin, *Optical Clearing of Tissues and Blood* (SPIE Press, Bellingham, 2006)
2. I. Carneiro, S. Carvalho, V. Silva, R. Henrique, L. Oliveira, V.V. Tuchin, Kinetics of optical properties of human colorectal tissues during optical clearing: a comparative study between normal and pathological tissues. J. Biomed. Opt. **23**(12), 121620 (2018)
3. A.Y. Sdobnov, M.E. Darvin, E.A. Genina, A.N. Bashkatov, J. Lademann, V.V. Tuchin, Recent progress in tissue clearing for spectroscopic application. Spectrochim. Acta A Mol. Biomol. Spectrosc. **197**, 216–229 (2018)
4. E.A. Genina, A.N. Bashkatov, V.V. Tuchin, Tissue optical immersion clearing. Expert Rev. Med. Devices **7**(6), 825–842 (2010)

5. D. Zhu, K.V. Larin, Q. Luo, V.V. Tuchin, Recent progress in tissue optical clearing. Laser Photon. Rev. **7**(5), 732–757 (2013)
6. L. Oliveira, M.I. Carvalho, E. Nogueira, V.V. Tuchin, Optical measurements of rat muscle samples under treatment with ethylene glycol and glucose. J. Innov. Opt. Health Sci. **6**(2), 1350012 (2013)
7. I. Carneiro, S. Carvalho, R. Henrique, L. Oliveira, V.V. Tuchin, Water content and scatterers dispersion evaluation in colorectal tissues. J. Biomed. Photon. Eng. **3**(4), 040301-1–040301-10 (2017)
8. D. R. Lide (ed.), *CRC Handbook of Chemistry and Physics* (CRC Press, Boca Raton, FL, 2005). Internet Version 2005, http://www.hbcpnetbase.com
9. https://www.accessdata.fda.gov/drugsatfda_docs/label/2017/018956s099lbl.pdf. Accessed 24 June 2019
10. A.Y. Sdobnov, M.E. Darvin, J. Schleusener, J. Lademann, V.V. Tuchin, Hydrogen bound water profiles in the skin influenced by optical clearing molecular agents – quantitative analysis using confocal Raman microscopy. J. Biophotonics **12**(5), e201800283 (2019)
11. S. Carvalho, N. Gueiral, E. Nogueira, R. Henrique, L. Oliveira, V.V. Tuchin, Glucose diffusion in colorectal mucosa: a comparative study between normal and cancer tissues. J. Biomed. Opt. **22**(9), 091506 (2017)
12. L. Oliveira, M.I. Carvalho, E. Nogueira, V.V. Tuchin, Diffusion characteristics of ethylene glycol in skeletal muscle. J. Biomed. Opt. **20**(5), 051019 (2015)
13. L. Oliveira, M.I. Carvalho, E. Nogueira, V.V. Tuchin, The characteristic time of glucose diffusion measured for muscle tissue at optical clearing. Laser Phys. **23**(7), 075606 (2013)
14. P. Wan, J. Zhu, J. Xu, Y. Li, T. Yu, D. Zhu, Evaluation of seven optical clearing methods in mouse brain. Neurophotonics **5**(3), 035007 (2018)
15. K. Tainaka, T.C. Murakami, E.A. Susaki, C. Shimizu, R. Saito, K. Takahashi, A. Hayashi-Takagi, H. Sekiya, Y. Arima, S. Nojima, M. Ikemura, T. Ushiku, Y. Shimizu, M. Murakami, K.F. Tanaka, M. Iino, H. Kasai, T. Sasaoka, K. Kobayashi, K. Miyazono, E. Morii, T. Isa, M. Fukayama, A. Kakita, H.R. Ueda, Chemical landscape for tissue clearing based on hydrophilic reagents. Cell Rep. **24**(8), 2196–2210 (2018)
16. https://refractiveindex.info. Accessed 3 Feb 2019
17. I.Z. Kozma, P. Krok, E. Riedle, Direct measurement of the group-velocity mismatch and derivation of the refractive-index dispersion for a variety of solvents in the ultraviolet. J. Opt. Soc. Am. B **22**(7), 1479–1485 (2005)
18. T.P. Otanicar, P.E. Phelan, J.S. Golden, Optical properties of liquids for direct absorption solar thermal energy systems. Sol. Energy **83**, 969–977 (2009)
19. R.D. Birkhoff, L.R. Painter, J.M. Heller Jr., Optical and dielectric functions of liquid glycerol from gas photoionization measurements. J. Chem. Phys. **69**(9), 4185–4188 (1978)
20. E. Sani, A. Dell'Oro, Optical constants of ethylene glycol over an extremely wide spectral range. Opt. Mater. **37**, 36–41 (2014)
21. I.Y. Yanina, E.N. Lazareva, V.V. Tuchin, Refractive index of adipose tissue and lipid droplet measured in wide spectral and temperature ranges. Appl. Opt. **57**(17), 4839–4848 (2018)
22. E.N. Lazareva, V.V. Tuchin, Measurement of refractive index of hemoglobin in the visible/NIR spectral range. J. Biomed. Opt. **23**(3), 035004 (2018)
23. E.N. Lazareva, V.V. Tuchin, Blood refractive index modelling in the visible and near infrared spectral regions. J. Biomed. Photon. Eng. **4**(1), 010503-1–010503-7 (2018)
24. I. Carneiro, S. Carvalho, R. Henrique, L. Oliveira, V.V. Tuchin, Simple multimodal optical technique for evaluation of free/bound water and dispersion of human liver tissue. J. Biomed. Opt. **22**(12), 125002 (2017)
25. I. Carneiro, S. Carvalho, R. Henrique, L. Oliveira, V.V. Tuchin, Optical properties of colorectal muscle in visible/NIR range, in *Proc SPIE 10685, Biophotonics: Photonic Solutions for Better Health Care VI*, (SPIE Press, Bellingham, 2018), p. 106853D. https://doi.org/10.1117/12.2306586

26. Y. Damestani, B. Melakeberhan, M.P. Rao, G. Aguilar, Optical clearing agent perfusion enhancement via combination of microneedle poration, heating and pneumatic pressure. Lasers Surg. Med. **46**(6), 488–498 (2014)

27. E.A. Genina, A.N. Bashkatov, A.A. Korobko, E.A. Zubkova, V.V. Tuchin, I. Yaroslavsky, G.B. Altshuler, Optical clearing of human skin: comparative study of permeability and dehydration of intact and photothermally perforated skin. J. Biomed. Opt. **13**(2), 021102 (2008)

28. J. Yoon, T. Son, E.-H. Choi, B. Choi, J.S. Nelson, B. Jung, Enhancement of optical skin clearing efficacy using a microneedle roller. J. Biomed. Opt. **13**(2), 021103 (2008)

29. J. Yoon, D. Park, T. Son, J. Seo, J.S. Nelson, B. Jung, A physical method to enhance transdermal delivery of a tissue optical clearing agent: combination of microneedling and sonophoresis. Lasers Surg. Med. **42**, 412–417 (2010)

30. C. Liu, Z. Zhi, V.V. Tuchin, Q. Luo, D. Zhu, Enhancement of skin optical clearing efficacy using photo-irradiation. Lasers Surg. Med. **42**, 132–140 (2010)

31. O.F. Stumpp, A.J. Welch, T.E. Milner, J. Neev, Enhancement of transepidermal skin clearing agent delivery using a 980 nm diode laser. Lasers Surg. Med. **37**, 278–285 (2005)

32. S. Karma, J. Homan, C. Stoianovici, B. Choi, Enhanced fluorescence imaging with DMSO-mediated optical clearing. J. Innov. Opt. Health Sci. **3**(3), 153–158 (2010)

33. E.A. Genina, A.N. Bashkatov, E.A. Kolesnikova, M.V. Basco, G.S. Terentyuk, V.V. Tuchin, Optical coherence tomography monitoring of enhanced skin optical clearing in rats in vivo. J. Biomed. Opt. **19**(2), 021109 (2014)

34. J. Jiang, R.K. Wang, Comparing the synergistic effects of oleic acid and dimethyl sulfoxide as vehicles for optical clearing of skin tissue in vitro. Phys. Med. Biol. **49**, 5283–5294 (2004)

35. J. Jiang, R.K. Wang, How different molarities of oleic acid as enhancer exert its effect on optical clearing of skin tissue in vitro. J. X-Ray Sci. Technol. **13**, 149–159 (2005)

36. A. Liopo, R. Su, D.A. Tsyboulsky, A. Oraevsky, Optical clearing of skin enhanced with hyaluronic acid for increased contrast of optoacoustic imaging. J. Biomed. Opt. **21**(8), 081208 (2016)

37. Z. Zhi, Z. Han, Q. Luo, D. Zhu, Improve optical clearing of skin in vitro with propylene glycol as a penetration enhancer. J. Innov. Opt. Health Sci. **2**(3), 269–278 (2009)

38. P. Karande, A. Jain, K. Ergun, V. Kispersky, S. Mitragotri, Design principles of chemical penetration enhancers for transdermal drug delivery. Proc. Natl. Acad. Sci. U. S. A. **102**, 4688–4693 (2005)

39. L. Oliveira, A. Lage, M. Pais Clemente, V.V. Tuchin, Rat muscle opacity decrease due to the osmosis of a simple mixture. J. Biomed. Opt. **15**(5), 055004 (2010)

40. M.H. Khan, B. Choi, S. Chess, K.M. Kelly, J. McCullough, J.S. Nelson, Optical clearing of in vivo human skin: implications for light-based diagnostic imaging and therapeutics. Lasers Surg. Med. **34**, 83–85 (2004)

41. https://preparatorychemistry.com/Bishop_supersaturated.htm. Accessed 4 July 2019

42. http://edge.rit.edu/edge/P13051/public/ResearchNotes/refractive index glycerin water.pdf. Accessed 10 Feb 2019

43. https://www.mt.com/us/en/home/supportive_content/concentration-tables-ana/Ethylene_Glycol_re_e.html. Accessed 10 Feb 2019

44. L.M. Oliveira, *The Effect of Optical Clearing in the Optical Properties of Skeletal Muscle*, PhD-thesis, FEUP Edições, 2014

45. R.M.H. Verbeeck, H.P. Thun, F. Veerbeek, Refractive index of the propylene glycol-water system from 15 to 50 °C. Bull. Soc. Chim. Belg. **85**(8), 531–534 (1976)

46. http://www.refractometer.pl/refraction-datasheet-sucrose. Accessed 12 Feb 2019

Chapter 4
Major Optical Clearing Mechanisms

4.1 Nature of Strong Light Scattering in Biological Tissues

In Chap. 1, we have presented some examples of biological tissues that have strong scattering properties and explained that such high scattering is created in a significant part by the refractive index (RI) mismatch between tissue components and fluids. Biological tissues are heterogeneous materials that contain cells and their organelles, like mitochondria, nucleus, lipid droplets, and phospholipid membranes, as well as protein fibers and globules, which are surrounded by tissue fluids like the cytoplasm and interstitial fluid (ISF) [1]. These cell and tissue structures are generally designated as scatterers, and they present higher RI values than cell or tissue fluids [2]. A measure of light scattering in a tissue can be calculated through the evaluation of the relative RI [3, 4]:

$$m(\lambda) = \frac{n_{\text{scat}}(\lambda)}{n_{\text{ISF}}(\lambda)}.$$

(4.1)

In Eq. (4.1), we have the ratio between the RI of tissue scatterers (n_{scat}) and the RI of the ISF of the tissue (n_{ISF}). Since both n_{scat} and n_{ISF} depend on wavelength, the ratio calculated by Eq. (4.1) also depends on the wavelength, but in general m is referred to as the reference wavelength of the Abbe refractometer (589.6 nm) [4].

The RI of various tissue components has been evaluated or calculated, and estimations of m for those tissues can be determined. The ISF of biological tissues is mainly water, where a small amount of salts and organic compounds are dissolved [4–6], meaning that the RI of ISF at 589.6 nm and 20 °C ranges between 1.35 and 1.37 [4].

The RI of normally hydrated collagen fibers in scleral tissues has been estimated as 1.474 [7], which gives an m ranging from 1.08 to 1.09. Dry skeletal muscle fibers have been estimated to have a RI of 1.584 [6], while at normal hydration the RI drops to 1.53 [1, 5], meaning that in natural conditions, m ranges between approximately

© The Author(s), under exclusive license to Springer Nature Switzerland AG 2019
L. M. C. Oliveira, V. V. Tuchin, *The Optical Clearing Method*,
SpringerBriefs in Physics, https://doi.org/10.1007/978-3-030-33055-2_4

1.12 and 1.13. For skin melanin, the RI is even higher, having a value of 1.6 [4], meaning that the m value ranges between 1.16 and 1.19. For any particular tissue, m will always be above 1, and its magnitude depends both on the RI of the dry component and its hydration in normal state.

Considering as example a muscle cell, at a local interface between the sarcoplasm and a muscle fiber, a photon will be scattered due to the RI mismatch. Since in a biological tissue there are many local interfaces, the scattering probability of each photon is high. To minimize, or even to eliminate light scattering in biological tissues, m should reduce to values just above 1 or even 1 exactly. The only way to do it is to change the RI of the tissue fluids, raising it to approximate or match the RI of tissue scatterers. The optical immersion clearing method is an efficient way to perform this matching, but it relies on some mechanisms that need to be understood to optimize treatments. The following sections explain these mechanisms.

4.2 Water and OCA Fluxes and the Clearing Mechanisms

To explain the water and OCA exchange between the tissue and the treating solution and relate those fluxes with the clearing mechanisms, we will consider an ex vivo slab-form tissue sample, which is immersed in a solution that contains a hyperosmotic agent, such as glucose or glycerol. The volume of the solution should be considerably higher than sample volume, so that the water and agent fluxes occur continuously.

Once the tissue sample is immersed in the solution, the optical clearing agent (OCA) creates an osmotic pressure over the sample, stimulating the water flux from the ISF to the outside. This water flux out leads to sample thickness decrease and to the approximation of tissue scatterers, forming a more compact and more organized packing inside [4, 8–10]. Mechanical tissue compressing or stretching performs the same stimulation to remove water from the compressed or stretched area [4]. As a consequence of a denser scatterer distribution inside, an increase in the scattering and absorption coefficients is expected, but due to a less sample thickness and better ordering of scatterers, which leads to constructive interference of the scattered waves in the propagation direction of the incident beam, the sample becomes more transparent [4, 8]. The water flux out of the tissue is designated as the dehydration mechanism [8].

At early stage of treatment, the OCA molecules not only provide the osmotic pressure to induce the water flux out. Since these OCAs are hyperosmotic, their molecules also diffuse into the tissue. With the water molecules flowing out, the early stage diffusion of OCA molecules consists only of their interaction with the superficial layers of the tissue. As water continues to flow out, it becomes easier for the OCA molecules to penetrate deeper into all the ISF areas of the tissue. Both fluxes (water going out and OCA going in) occur simultaneously, and means to discriminate and characterize them are necessary.

Since the RI of the OCA is higher than the RI of water, its diffusion into the ISF will provide an increase in the mean RI of the ISF, turning it more approximated to the RI of tissue scatterers. The OCA diffusion into the ISF provides the RI matching mechanism [8, 10, 11]. As a result of these mechanisms, an increase in tissue transparency is created [12], and better contrast images from deeper layers inside the tissue are obtained [13]. Both the tissue dehydration and RI matching mechanism are reversible. For in vivo tissues, the water in adjacent tissues will eventually flow into the treated tissue, washing the agent out. For ex vivo tissue, assisted tissue rehydration can be performed to revert the clearing effect.

Various tissues were submitted to studies during optical clearing (OC) treatments with glycerol [14–16], glucose [8, 12, 17], omnipaque™ [16], dimethyl sulfoxide [18], or ethylene glycol (EG) [8] with the objective of clarifying and characterizing the OC mechanisms. The investigation of the dehydration mechanism has also been made with application of mechanical forces and temperature variation methods to remove water from the areas of tissues under study [12, 19–22]. All these studies have reported that as a consequence of the water loss, tissue transparency increases due to the combination of smaller tissue sample with better scatterer ordering inside.

During tissue dehydration, the volume fractions for tissue scatterers (f_s) and ISF ($f_{ISF} = 1 - f_s$, see Eq. (1.16)) will change due to the water loss. Assuming that only interstitial water is lost during dehydration, the absolute scatterers' volume (V_s) will remain unchanged. This means that for a slab sample (see Fig. 2.1), the variations in sample volume (V) are driven by sample thickness variations (d). At early OC treatment, the dehydration mechanism is fast and stronger than the RI matching mechanism. Such fact leads to a strong decrease in d during dehydration, which implies also a strong V decrease ($V \propto d$). Since f_s is calculated as the ratio between V_s and V, with the decrease in d, f_s will increase. In fact, if all the interstitial water was lost, f_s would reach 1. In opposition, while f_s increases, f_{ISF} decreases in the same proportion [1, 3, 23].

It has been reported that during tissue dehydration and for a dense distribution of scattering particles, the reduced scattering coefficient (μ'_s) is proportional to the product of the volume fractions of tissue components [19]:

$$\mu'_s = \frac{f_s(1 - f_s)}{V_p} \sigma'_s, \tag{4.2}$$

where V_p is the volume of a single scattering particle and σ'_s represents the reduced scattering cross section (in cm^2) for a single scattering particle, which depends on the wavelength of light in vacuum, on the scattering particle's radius and RI, and on the RI of the surrounding medium. Such heuristic model shows that μ'_s will be zero both for $f_s = 0$ and for $f_s = 1$. We can represent μ'_s as a function of f_s to show that such model has parabolic dependence. Figure 4.1 shows that for any arbitrary V_p and σ'_s values used in Eq. (4.2), μ'_s shows a maximum for $f_s = 0.5$.

Fig. 4.1 Relationship
between μ'_s and f_s, provided
that the light wavelength in
vacuum, the scattering
particle radius and its RI,
and the RI of surrounded
medium remain unchanged

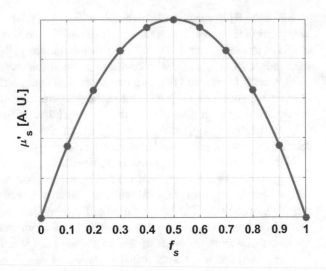

Fig. 4.1 Relationship between μ'_s and f_s, provided that the light wavelength in vacuum, the scattering particle radius and its RI, and the RI of surrounded medium remain unchanged

Equation (4.1) is useful for examining the dehydration mechanism of OC, since during this mechanism, the scattering particle density and ordering inside the tissue change due to the water loss [19].

During the RI matching mechanism, the OCA may change the scattering properties of tissue by changing scattering particle size, shape, packing density, its RI, and the RI of surrounding medium. Such changes depend on the kinetics observed for the water flux out and OCA flux in to the tissue and are to be accounted in σ'_s [19].

The RI matching mechanism has also been studied, and good results were obtained. Skin treatments with some particular OCAs in different concentrations and osmolarities showed different clearing potentials [24]. Aqueous solutions containing different concentrations of EG and glucose were used to treat skeletal muscle and result in differentiated magnitudes and transparency kinetics over 30 min of treatment [10, 25]. When comparing to normal tissues, tissues with enhanced permeability show different kinetics for both OC mechanisms [26], where the increase in transparency is better when the permeability is enhanced. Other studies have demonstrated that increased transmittance can also be obtained in treatments where alcohols are used as OCA diffusion enhancers [6, 27]. Sample thickness was also measured in these studies before and after the treatment, showing a small global variation.

Multiphoton microscopy has been used to generate second harmonic generation (SHG) images from ex vivo skin samples under treatment with glycerol solutions [28]. Such images have showed dissociation of collagen fibers in the extracellular skin matrix during OC treatments, but their reassembly was also observed when, after treatment, the skin samples were rehydrated in saline to wash out the glycerol. Yeh's group also studied the reversibility of protein dissociation in the skin when treated with sorbitol, EG, and glycerol [29]. This study demonstrated that the OC potential of an OCA is related to its protein solubility capability, suggesting though that protein solubility is also an OC mechanism. This study has also demonstrated that hyperosmolarity of an OCA relative to tissue forces sample dehydration. Hydrogen bond bridge formation disrupts the collagen hydration layer and facilitates water replacement by the OCA to perform RI matching [30]. Some studies, performed with in vivo rat dorsal skin, have showed no collagen dissociation or fracturation at used OCA concentrations [31]. This study showed instead that skin thickness and collagen diameter have decrease as a result of water loss. V. Hovhannisyan et al. have used a time-lapse multiphoton microscopy technique to study the OC mechanisms [32]. Animal skin and tendon tissues were treated with high concentrated glycerol solutions, and results show that collagen suffered fast dehydration and shrinkage, while glycerol penetration was slower than dehydration and accompanied with sample swelling.

Collagen dissociation was observed in some ex vivo studies, but the tissue dehydration and the RI matching mechanisms are always observed during all OC treatments. These two mechanisms are the main driving forces to create tissue transparency through the decrease of light scattering in the OC treatments. Since the two main OC mechanisms are driven by the water flux out and the OCA flux in, it is necessary to find a way to individualize these mechanisms and fluxes, by establishing a quantitative and descriptive characterization for each one. A method based on collimated transmittance (T_c) and thickness measurements performed during treatment of a tissue with solutions of an OCA with particular concentrations allows for such characterization. Such method is described in the following section.

4.3   Characterization of the Water and OCA Fluxes and Mechanism Discrimination

In the previous section, we have associated the dehydration mechanism with the water flux out from the tissue and the RI matching mechanism with the OCA flux into the tissue. This means that if we can evaluate and characterize both these fluxes individually, we can discriminate and characterize the two OC mechanisms. Such discrimination and characterization can be made through thickness and T_c measurements from ex vivo tissue samples, as reported in literature [4, 33]. Such method uses the T_c measurements to estimate the diffusion time τ, both for water and for OCA, and thickness measurements are used further to calculate the diffusion coefficient, D, also both for water and OCA. Such calculations are made through Eqs. (2.4) and (2.7), which we reproduce here:

$$T_c(\lambda, t) = \frac{C_a(t)}{C_{a0}} \cong \left[1 - \exp\left(-\frac{t}{\tau}\right)\right], \qquad (4.3)$$

$$D_a = \frac{d^2}{\pi^2 \times \tau}. \qquad (4.4)$$

By fitting the T_c time dependencies for individual wavelengths within a wavelength range where no absorption bands occur with Eq. (4.3), individual τ values are estimated. The mean τ can then be calculated, and using it in Eq. (4.4), D can be calculated [34]. Equation (4.4) is used for OCA and water fluxes through both slab surfaces. In case that these fluxes occur only through one slab surface, Eq. (4.5) should be used [4].

$$\tau = \frac{4d^2}{\pi^2 D_a}. \tag{4.5}$$

Such calculation/estimation procedure can be performed for an ex vivo tissue under treatment with aqueous solutions containing any OCA concentration. In a treatment with a particular solution, the estimated τ and D values represent the effective diffusion time and diffusion coefficient for the combined flux of water going out and OCA going in. If we want to obtain these values for the particular cases of unique water flux out and unique OCA flux in, we must select the OCA concentrations in the treating solution carefully [35]. If the mobile water content in the tissue is known, to obtain the unique OCA flux in, such water content must be matched by the water content in the treating solution. This way, due to the water balance between the tissue and the solution, only the OCA flows in. To obtain a unique water flux out, the OCA concentration in solution must be high. That way, due to the strong OCA concentration in the treating solution, a strong osmotic pressure is created over the tissue, which stimulates the water flux out within the first few minutes of treatment [10, 25]. These particular cases of unique water flux out and unique OCA flux in are presented in Fig. 4.2a, b for skeletal muscle treated with 40% and 60% fructose solutions. Figure 4.2c presents the OC efficiency of three sugars in ex vivo rat skin as a function of sugar concentration in the treating solution. The OC efficiency is represented with the designation RSR (reduced scattering ratio), and it was calculated for fructose (a monosaccharide) and sucrose and maltose (two disaccharides) as follows [36]:

$$RSR = \frac{\mu'_{before}}{\mu'_{after}}. \tag{4.6}$$

The calculations made with Eq. (4.6) were performed at $\lambda = 635$ nm using the reduced scattering coefficient before (μ'_{before}) and after (μ'_{after}) the treatment.

For the skin treatments with sugar solutions, presented in Fig. 4.2c, we observe that OC efficiency increases linearly with sugar concentration in the treating solution. Such behavior is observed for all sugars and in different ranges of sugar concentrations. For the case of fructose, the slope of the linear increase decreases as the range of concentrations increases from 1–2 M to 1–3 M and then to 1–6 M. Sucrose shows opposite behavior, since the slope of the linear increase in OC efficiency decreases from the 1–2 M to the 1–3 M concentration range [36]. This same study also evaluated the variations in the optical coherence tomography (OCT)

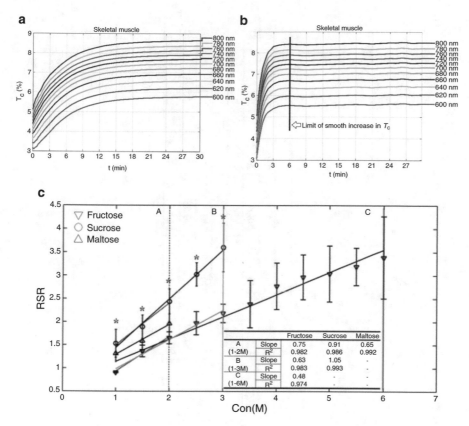

Fig. 4.2 T_c time dependencies of skeletal muscle under treatment with 40%-fructose (**a**) and 60%-fructose (**b**) and OC efficiency as a function of sugar concentration (**c**). Figure (**c**) is reprinted with permission from Ref. [36]

signal during treatment at skin depths of 300 and 400 µm. It was observed that the signal intensity also increases linearly during treatment at both skin depths, but the increase for treatments with saturated sucrose shows higher slope than the one observed for saturated fructose [36].

In Fig. 4.2a we see a smooth increase in T_c during all treatment time for all wavelengths. This means that due to the water balance between the treating solution and tissue mobile water, a unique fructose flow into the tissue is observed. In Fig. 4.2b, due to the high fructose concentration in the treating solution (60%), we see a smooth increase in T_c within the first 6 min of treatment. This means that the unique water flux out from the tissue is responsible for this initial smooth increase in T_c. By fitting each curve in graphs of Fig. 4.2 with a curve described by Eq. (4.3), we can obtain the correspondent τ values and then calculate the mean for each treatment. Considering the treatments presented in Fig. 4.2, those values are presented in Table 4.1.

Table 4.1 Estimated τ values for the treatments of muscle with fructose solutions

Treatment with 40%-fructose

λ (nm)	600	620	640	660	680	700	720	740	760	780	800
τ (s)	338.5	332.4	331.7	325.1	320.3	312.7	305.7	298.8	292.8	287.2	281.1
		Mean τ (s)		311.48		SD (s)		19.74			

Treatment with 60%-fructose

λ (nm)	600	620	640	660	680	700	720	740	760	780	800
τ (s)	58.34	59.27	59.73	59.77	58.93	57.90	57.34	56.58	55.99	55.41	55.31
		Mean τ (s)		57.69		SD (s)		1.68			

The diffusion coefficients for water and fructose in skeletal muscle were obtained using the mean τ values presented in Table 4.1 and the sample thickness at those times of treatment in Eq. (4.4): $D_{water} = 2.9 \times 10^{-6}$ cm^2/s, $D_{fructose} = 5.7 \times 10^{-7}$ cm^2/s [33].

Both the τ and D values obtained for water and fructose characterize the water flux out and the fructose flux into the muscle and also the dehydration and the RI matching mechanisms. Those values are individual for each tissue when treated with a particular OCA.

The method used to evaluate the diffusion properties of OCAs in biological tissues has other applications, and it will be described in detail in Chap. 6. Due to the nature of these measurements, such method can only be applied with ex vivo tissues, but similar procedure can be developed for in vivo situation. If instead of T_c we use diffuse reflectance (R_d) measurements, the evaluation of τ can be made from in vivo tissues under treatment. In vivo tissue thickness kinetics can be evaluated during treatment by OCT or confocal microscopy measurements [18, 37].

References

1. L. Oliveira, M.I. Carvalho, E. Nogueira, V.V. Tuchin, Skeletal muscle dispersion (400–1000 nm) and kinetics at optical clearing. J. Biophotonics **11**(1), e201700094 (2018)
2. V. Backman, R. Gurjar, K. Badizadegan, I. Itzkan, R.R. Dasari, L.T. Perelman, M.S. Feld, Polarized light scattering spectroscopy for quantitative measurement of epithelial cellular structures in situ. IEEE J. Sel. Top. Quant. Electron. **5**(4), 1019 (1999)
3. I. Carneiro, S. Carvalho, V. Silva, R. Henrique, L. Oliveira, V.V. Tuchin, Kinetics of optical properties of human colorectal tissues during optical clearing: a comparative study between normal and pathological tissues. J. Biomed. Opt. **23**(12), 121620 (2018)
4. V.V. Tuchin, *Optical Clearing of Tissues and Blood* (SPIE Press, Bellingham, 2006)
5. L. Oliveira, A. Lage, M. Pais Clemente, V.V. Tuchin, Optical characterization and composition of abdominal wall muscle from rat. Opt. Lasers Eng. **47**, 667–672 (2009)
6. L. Oliveira, A. Lage, M. Pais Clemente, V.V. Tuchin, Rat muscle opacity decrease due to the osmosis of a simple mixture. J. Biomed. Opt. **15**(5), 055004 (2010)
7. A.N. Bashkatov, E.A. Genina, V.I. Kochubey, V.V. Tuchin, Estimation of wavelength dependence of refractive index of collagen fibers of scleral tissue, in *Controlling Tissue Optical Properties: Applications in Clinical Study*, Proc. of SPIE, ed. by V. V. Tuchin, vol. 4162, (SPIE Press, Bellingham, 2000), pp. 265–268
8. L. Oliveira, M.I. Carvalho, E. Nogueira, V.V. Tuchin, Optical clearing mechanisms characterization in muscle. J. Innov. Opt. Health Sci. **9**(5), 1650035 (2016)
9. V.V. Tuchin, *Tissue Optics: Light Scattering Methods and Instruments for Medical Diagnosis*, 3rd edn. (SPIE Press, Bellingham, 2015)
10. L. Oliveira, M.I. Carvalho, E. Nogueira, V.V. Tuchin, Diffusion characteristics of ethylene glycol in skeletal muscle. J. Biomed. Opt. **20**(5), 051019 (2015)
11. J. Hirshburg, B. Choi, J.S. Nelson, A.T. Yeh, Correlation between collagen solubility and skin optical clearing using sugars. Laser. Surg. Med. **39**, 140–144 (2007)
12. E.A. Genina, A.N. Bashkatov, V.V. Tuchin, Tissue optical immersion clearing. Expert Rev. Med. Dev. **7**(6), 825–842 (2010)
13. R. Cicchi, D. Sampson, D. Massi, F.S. Pavone, Contrast and depth enhancement in two-photon microscopy of human skin ex vivo by use of optical clearing agents. Opt. Express **13**(7), 2337–2344 (2005)

14. G. Vargas, E.K. Chan, J.K. Barton, H.G. Rylander, A.J. Welch, Use of an agent to reduce scattering in skin. Laser. Surg. Med. **24**, 133–141 (1999)
15. G. Vargas, J.K. Barton, A.J. Welch, Use of hyperosmotic chemical agent to improve the laser treatment of cutaneous vascular lesions. J. Biomed. Opt. **13**(2), 021114 (2008)
16. A.Y. Sdobnov, M.E. Darvin, J. Schleusener, J. Lademan, V.V. Tuchin, Hydrogen bound water profiles in the skin influenced by optical clearing molecular agents – quantitative analysis using confocal Raman microscopy. J. Biophotonics **12**(5), e201800283 (2019)
17. H. Zheng, J. Wang, Q. Ye, Z. Deng, J. Mei, W. Zhou, C. Zhang, J. Tian, Study on the refractive index matching effect of ultrasound on optical clearing of bio-tissues based on the derivative total reflection method. Biomed. Opt. Express **5**(10), 3482–3493 (2014)
18. D. Zhu, K.V. Larin, Q. Luo, V.V. Tuchin, Recent progress in tissue optical clearing. Laser Photon. Rev. **7**(5), 732–757 (2013)
19. C.G. Rylander, O.F. Stumpp, T.E. Milner, N.J. Kemp, J.M. Mendenhall, K.R. Diller, A.J. Welch, Dehydration mechanism of optical clearing in tissue. J. Biomed. Opt. **11**(4), 041117 (2006)
20. T. Yu, X. Wen, V.V. Tuchin, Q. Luo, D. Zhu, Quantitative analysis of dehydration in porcine skin for assessing mechanism of optical clearing. J. Biomed. Opt. **16**(9), 095002 (2011)
21. A.A. Gurjarpadhye, W.C. Vogt, Y. Liu, C.G. Rylander, Effect of localized mechanical indentation on skin water content evaluated using OCT. Int. J. Biomed. Imaging **2011**, 817250 (2011)
22. Y. Tanaka, A. Kubota, M. Yamato, T. Okano, K. Nishida, Irreversible optical clearing of sclera by dehydration and cross-linking. Biomaterials **32**, 1080–1090 (2011)
23. I. Carneiro, S. Carvalho, R. Henrique, L. Oliveira, V.V. Tuchin, Kinetics of optical properties of colorectal muscle during optical clearing. IEEE J. Sel. Top. Quant. Electron **25**(1), 7200608 (2019)
24. B. Choi, L. Tsu, E. Chen, T.S. Ishak, S.M. Iskhandar, S. Chess, J.S. Nelson, Determination of chemical agent optical clearing potential using in vitro human skin. Lasers Surg. Med. **36**, 72–75 (2005)
25. L. Oliveira, M.I. Carvalho, E.M. Nogueira, V.V. Tuchin, The characteristic time of glucose diffusion measured for muscle tissue at optical clearing. Laser Phys. **23**, 075606 (2013)
26. E.A. Genina, A.N. Bashkatov, A.A. Korobko, E.A. Zubkova, V.V. Tuchin, I. Yaroslavsky, G.B. Altshuler, Optical clearing of human skin: comparative study of permeability and dehydration of intact and photothermally perforated skin. J. Biomed. Opt. **13**(2), 021102 (2008)
27. Z. Mao, D. Zhu, Y. Hu, X. Wen, Z. Han, Influence of alcohols on the optical clearing effect of skin in vitro. J. Biomed. Opt. **13**(2), 021104 (2008)
28. A. Yeh, B. Choi, J.S. Nelson, B.J. Tromberg, Reversible dissociation of collagen in tissues. J. Invest. Dermatol. **121**, 1332–1335 (2003)
29. J. Hirshburg, B. Choi, J.S. Nelson, A.T. Yeh, Collagen solubility correlates with skin optical clearing. J. Biomed. Opt. **11**, 040501 (2006)
30. J. Hirshburg, K.M. Ravikumar, W. Hwang, A.T. Yeh, Molecular basis for optical clearing of collagenous tissues. J. Biomed. Opt. **15**(5), 055002 (2010)
31. X. Wen, Z. Mao, Z. Han, V.V. Tuchin, D. Zhu, In vivo skin optical clearing by glycerol solutions: mechanism. J. Biophotonics **3**, 44–52 (2010)
32. V. Hovhannisyan, P.-S. Hu, S.-J. Chen, C.-S. Kim, C.-Y. Dong, Elucidation of the mechanisms of optical clearing in collagen tissue with multiphoton imaging. J. Biomed. Opt. **18**(4), 046004 (2013)
33. I. Carneiro, S. Carvalho, R. Henrique, L. Oliveira, V.V. Tuchin, A robust ex vivo method to evaluate the diffusion properties of agents in biological tissues. J. Biophotonics **12**(4), e201800333 (2019)
34. I. Carneiro, S. Carvalho, R. Henrique, L. Oliveira, V.V. Tuchin, Simple multimodal optical technique for evaluation of free/bound water and dispersion of human liver tissue. J. Biomed. Opt. **22**(12), 125002 (2017)

35. S. Carvalho, N. Gueiral, E. Nogueira, R. Henrique, L. Oliveira, V.V. Tuchin, Glucose diffusion in colorectal mucosa – a comparative study between normal and cancer tissues. J. Biomed. Opt. **22**(9), 091506 (2017)
36. W. Feng, R. Shi, N. Ma, D.K. Tuchina, V.V. Tuchin, D. Zhu, Skin optical clearing potential of disaccharides. J. Biomed. Opt. **21**(8), 081207 (2016)
37. K. Schilling, V. Janve, Y. Gao, I. Stepniewska, B.A. Landman, A.W. Anderson, Comparison of 3D orientation distribution functions measured with confocal microscopy and diffusion MRI. NeuroImage **129**, 185–197 (2016)

Chapter 5
Measurements During Optical Clearing

5.1 Ex Vivo Measurements

Various measurements can be made for ex vivo biological tissues under optical clearing (OC) treatments. The purpose of those measurements can be diverse. From one hand, those measurements can be used to calculate the clearing efficiency of an agent, to calculate the diffusion properties of the agent, or, on the other hand, to estimate the kinetics of the optical properties of the tissue under treatment.

In the following subsections, we describe various measurements that can be made for ex vivo tissues, referring to their purpose, and in the following section, we will discuss the in vivo measurements that can be made during OC treatments.

5.1.1 Collimated Transmittance

Collimated transmittance (T_c) is a particular measurement that allows for gathering various information and to perform several calculations related to OC treatments. The T_c data is sensitive to changes that occur during treatments, but due to the nature of these measurements and to avoid invasive procedures, usually they are only made with ex vivo samples.

The T_c for a sample with thickness d is defined by the Bouguer-Beer-Lambert equation [1–7]:

$$T_c = \frac{I(d)}{I_0} = \exp\left(-\mu_t d\right), \tag{5.1}$$

where I_0 and $I(d)$ are the intensities of the incident and detected light across the sample, respectively, and $\mu_t = \mu_a + \mu_s$ is the attenuation coefficient. To perform such measurements, a particular setup, like the one represented in Fig. 5.1, is necessary.

© The Author(s), under exclusive license to Springer Nature Switzerland AG 2019
L. M. C. Oliveira, V. V. Tuchin, *The Optical Clearing Method*,
SpringerBriefs in Physics, https://doi.org/10.1007/978-3-030-33055-2_5

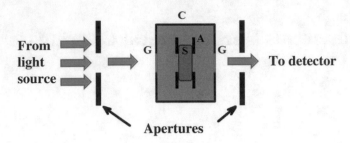

Fig. 5.1 Typical setup to measure T_c

In Fig. 5.1 the sample (S) is placed inside a cuvette (C), and a set of wires (A) fix the sample in measuring position. A collimated beam that comes from the light source passes through an aperture to reduce its diameter to 1 mm. This beam enters the cuvette through a glass window (G) and crosses the sample, exiting the cuvette through another glass window. An additional aperture is placed on the detector side of the cuvette to guarantee that only collimated light is sent to the detector. The light source can be a laser and the detector a photodetector, but to get spectral properties of OC, it is more common to use such setup with a broadband lamp and a spectrometer to acquire the collimated transmittance spectrum ($T_c(\lambda)$) of samples.

The setup presented in Fig. 5.1 can be used to measure the T_c spectra from a sample under OC treatment. In that case, the cuvette is filled with a solution containing an active OCA, and the spectra measured during treatment allow one to create a 3D kinetics for T_c, which is showing a set of spectra in time domain. Such data is of great interest for various applications. One example was given in Chap. 4, where T_c kinetics for various wavelengths were used to calculate the diffusion time (τ) values for fructose and water in skeletal muscle [8]. Similar studies can be performed to evaluate the efficiency of certain OCAs with different osmolarities [6, 8]. Some of these studies were used to discriminate between normal and pathological tissues through the estimation of different mobile water content [9, 10]. Studies to calculate the refractive index (RI) kinetics, based on T_c and thickness measurements, have already been performed [11, 12] and demonstrate the occurrence of the RI matching mechanism. Similar studies that use also T_c and thickness measurements to calculate the kinetics of the optical properties of biological tissues under OC treatments [7, 12] have also demonstrated a smooth decrease in the scattering and reduced scattering coefficients and an increase in the g-factor.

Similar to analogous measurements made from natural samples [13–18], data from T_c spectra can be used on inverse Monte Carlo (IMC) [19] or inverse adding-doubling (IAD) [20] simulations to estimate the kinetics of the optical properties of the tissue under treatment.

All these studies have proven valuable for gathering information relative to OC treatments. Further studies can be made with similar objectives for other research fields. As an example, the estimation of the diffusion properties of cryogenic agents for organ or meat preserving can be of interest. In such particular applications, T_c measurements can be applied, since they will be made with ex vivo tissue samples. Other complementary studies can be made also with ex vivo tissues to estimate the mobile water in tissues where it is not known or to discriminate other types of pathologies. The estimation of the diffusion properties of poisons, drugs, skin oils, or creams is of great interest.

A particular study where T_c measurements might prove valuable is the evaluation of proteins and DNA solubility during OC treatments. It has been described that different proteins and DNA contents are found in normal and pathological tissues [21]. The evaluation of such difference and of differentiated solubility during OC treatments can be used as a basis for future diagnosis procedure.

5.1.2   Total Transmittance

Total transmittance (T_t) measurements are in general made with the objective of estimating the optical properties of a sample through IMC [19] or IAD [20] simulations. T_t is the sum of T_c and diffuse transmittance (T_d). Once again, due to the nature of these measurements and since invasive measurements are not desired during OC treatments, T_t measurements are not suitable for in vivo studies. Regarding other biological samples, like meat, fruits, etc., the measurement of T_t during OC treatments can be of great interest to evaluate OC efficiency, for estimation of kinetics of optical properties during treatments, or to characterize the OC mechanisms, such as already described for T_c measurements.

During OC treatments, the g-factor is known to increase, and the scattering coefficient is known to decrease [12]. Due to these variations, it is expected that T_t increases during OC treatments. Some studies have been made that show the increase in T_t spectra during treatments. Figure 13b of Ref. [4] shows the increase of T_t spectra of eye sclera during treatment with Trazograph^{TM}-60.

The measurement of T_t spectra is made with the use of an integrating sphere, as represented in Fig. 5.2. The inside surface of the sphere is coated with $BaSO_4$, MgO, or Spectralon, which have nearly 100% reflectance over the entire optical spectrum [3, 22].

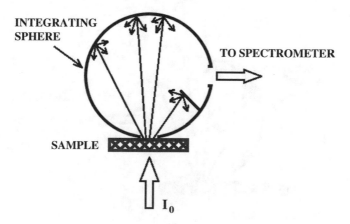

Fig. 5.2 Typical setup to measure T_t

According to Fig. 5.2, a beam from a broadband lamp crosses the sample and enters the integrating sphere through the sample port. Such beam is delivered through optical fiber cable and collimating lens. Multiple reflections (integration) occur inside the sphere before exiting through the exit port. The beam that exits the sphere is collected by an optical fiber cable to be delivered to the spectrometer, so that the T_t spectrum is registered. A baffle is placed between the sample and the exit ports of the sphere to avoid light to go directly from the sample to the exit port without being reflected at least once inside the sphere.

5.1.3   Total Reflectance

Similarly to the T_t measurements, total reflectance (R_t) measurements are also made with the use of an integrating sphere. R_t is the sum of specular reflectance (R_s) and diffuse reflectance (R_d). Both R_s and R_d can be measured from ex vivo or in vivo tissue samples, and they are both sensitive to changes made during OC treatments (see Figs. 9.40 and 9.43 of Ref. [5]). Due to the nature of the R_t measurements, they can also be made both from ex vivo or in vivo tissue samples.

These measurements can be useful during OC treatments, since due to the increase in the g-factor and decrease in the scattering coefficient, R_t should decrease over the time of treatment. If complimentary measurements are available, they can be used with R_t data on IMC or IAD simulations to estimate the kinetics of tissue's optical properties during treatment and though calculate OC efficiency or characterize the clearing mechanisms.

Figure 5.3 presents the typical setup to acquire R_t spectra.

According to Fig. 5.3, the beam from a broadband lamp enters the integrating sphere in a direction at 8° with the normal to the sample. This beam is reflected by the upper sample surface and then passes through multiple reflections at the inside

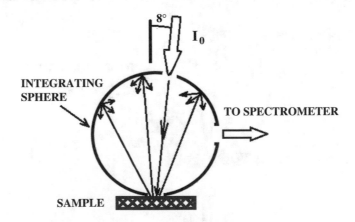

Fig. 5.3 Typical setup to measure R_t

Fig. 5.4 Goniometric setup
for RI measurements

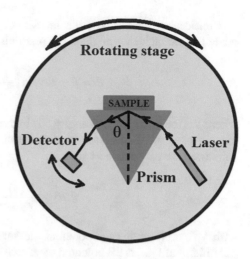

surface of the sphere (integration), before exiting to be delivered to the spectrometer. Normally, the light delivery into the integrating sphere is made via optical fiber cable and collimating lens. The R_t beam is collected from the detection port of the sphere with an optical fiber cable.

5.1.4 Refractive Index

The measurement of the refractive index (RI) of biological materials can be made using different setups. Considering ex vivo samples, the measurement of the RI can be made using conventional refractometers or using goniometric setups with different lasers.

Regarding conventional refractometers, the most common is the Abbe refractometer, which measures the RI at 589.6 nm. More recent refractometers can measure the RI at individual wavelengths from visible to infrared (see Fig. 3.1b). Such measurements at selective wavelengths are made by using interferential filters in the refractometer.

Considering goniometric RI measurements, a setup like the one presented in Fig. 5.4, which is based on the total internal reflection, should be used [23]. Such setup can be explored with various lasers at different wavelengths, with the objective of constructing the dispersion curve of the sample.

In the setup presented in Fig. 5.4, a laser beam is used to produce a reflected beam on the sample attached to the base of the prism. A voltmeter connected to a photodetector measures the reflected signal on the other side of the setup. The incident and reflected beams rotate around the setup, and the reflected signal is registered at various angles.

Considering Snell's law, the incidence angle at the air/prism interface (α) is related to the incidence angle at the prism/tissue interface (θ), according to [24]:

$$\theta = \beta - arcsin\left(\frac{1}{n_1} \times \sin(\alpha)\right), \tag{5.2}$$

where β is the prism internal angle (60° in most cases) and n_1 is the RI of the prism at the wavelength of the laser in use. Reflectance at the prism/tissue interface is calculated for each angle as [24]:

$$R(\theta) = \frac{V(\theta) - V_{\text{noise}}}{V_{\text{laser}} - V_{\text{noise}}}, \tag{5.3}$$

with $V(\theta)$ representing the potential measured at angle θ, V_{noise} is the background potential, and V_{laser} is the potential measured directly from the laser. A representation of the reflectance at the prism/tissue interface ($R(\theta)$) as a function of the incident angle (θ) is created, showing a behavior like the one presented in the curves of Fig. 5.5 for measurements made from human colorectal muscle at 668 nm [24].

Calculating the first derivative of the curves in Fig. 5.5, we obtain the curves presented in Fig. 5.6, which show a peak at a particular incidence angle. This is the critical angle (θ_c) of reflection for a particular laser between the prism and the tissue sample [24].

Averaging between the three values of θ_c obtained from the three curves, we can calculate the RI of the tissue sample at the wavelength of the laser. To perform such calculation, we use Eq. (5.4) [24].

$$n_t(\lambda) = n_1(\lambda) \times \sin(\theta_c) \tag{5.4}$$

In Eq. (5.4), $n_1(\lambda)$ represents the RI of the prism at the same wavelength of the laser. This procedure must be repeated for all lasers available, obtaining various RI values at different wavelengths.

After obtaining the RI values of a tissue at different wavelengths, such experimental data can be fitted with appropriate equations to calculate the corresponding dispersion curve. The usual equations to fit the RI values of biological materials are the Cauchy equation (Eq. 5.5), the Conrady equation (Eq. 5.6), and the Cornu equation (Eq. 5.7) [15, 25, 26]:

$$n(\lambda) = A + \frac{B}{\lambda^2} + \frac{C}{\lambda^4}, \tag{5.5}$$

$$n(\lambda) = A + \frac{B}{\lambda} + \frac{C}{\lambda^{3.5}}, \tag{5.6}$$

$$n(\lambda) = A + \frac{B}{(\lambda - C)}. \tag{5.7}$$

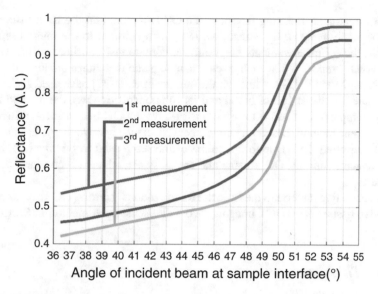

Fig. 5.5 Reflectance curves from human colorectal muscle at the tissue/prism interface measured with a red laser (668 nm laser)

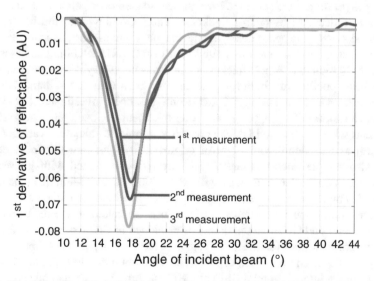

Fig. 5.6 First derivative of the reflectance curves presented in Fig. 5.5 (668 nm laser)

According to our calculations [11], the RI of a tissue under OC treatment should increase as a result of the RI matching between tissue scatterers and fluids. RI measurements during OC treatment are possible, but not with a conventional refractometer. Using an Abbe refractometer, the RI can be measured at discrete times of treatment, meaning that both the OC treatment and the RI measurements must be made in steps. If, instead, we use a setup similar to the one presented in Fig. 5.4, the RI measurements can be made continuously during treatment. It has been recently reported that such a setup can be made to acquire broadband spectral data almost instantly [27]. Although the authors of Ref. [27] have used such a setup to measure the dispersion of blood and its components ex vivo, the method is fast, and the setup can be used with in vivo tissues during OC treatment. A different RI measurement method, based on single fiber reflectance spectroscopy, has also been recently reported [28]. Due to the nature of the setup used in this method, in vivo measurements during OC treatments are possible, since spectra are acquired very fast.

An alternative method to measure the RI continuously during OC both for ex vivo or in vivo tissues is by OCT imaging. We will discuss such method in Sect. 5.2.

5.1.5 Thickness

According to our explanation for the OC mechanisms in Chap. 4, tissue samples suffer thickness variations during treatments. The water flow out that creates the tissue dehydration mechanism allows tissue scatterers to approach each other, leading to a decrease in sample thickness. The OCA flow into the tissue, on the other hand, reverts the decrease of sample thickness, forcing the tissue to swell as the agent molecules diffuse into the tissue to perform the RI matching mechanism. We have already seen that sample thickness kinetics is necessary if we want to calculate the diffusion coefficients, both for water and for OCA (see Eqs. 4.3 and 4.4).

Thickness variation during OC treatments has been investigated [29–32]. For comparison, skin thickness before and after treatment has been measured [29]. The treatments with glucose and DMSO created insignificant thickness variation of the skin [29] as treatments with glycerol originated a significant thickness decrease [30]. Also sample thickness decrease in myocardium treated with glycerol [31] and in the skin under treatment with PEG-300 and PEG-400 [32] have been observed. Thickness measurements made during treatment provide a great amount of information. Such measurements can be used to calculate the diffusion coefficients of water and OCAs in the tissue, as explained in Chap. 4. On the other hand, tissue thickness kinetics is useful to calculate the time dependence for the volume fractions of tissue components [11]. Such data are important to estimate the kinetics of the RI of the interstitial fluid (ISF) of the tissue during treatment and show that the RI matching mechanism occurs [12].

In some of previous studies, the kinetics of sample thickness were measured for tissues like skeletal muscle [6, 11, 33, 34] or colorectal tissues [7–9, 12, 15, 24] under treatment with various agents. The measured kinetics of tissue thickness was also used to calculate the time dependence of the optical properties of colorectal tissues, showing that the scattering coefficient decreases and the g-factor increases [7, 12].

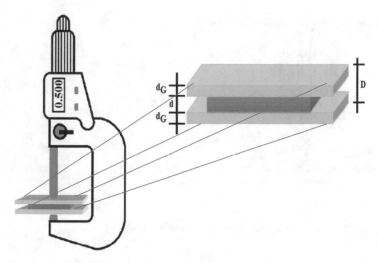

Fig. 5.7 Tissue sample thickness measuring setup

To measure thickness variation during OC treatments from ex vivo tissue samples, a setup like the one presented in Fig. 5.7 is commonly used.

To perform the thickness measurements during an OC treatment, a slab-form sample is initially placed inside two microscope glasses with known thickness, d_G. The solution that contains an OCA is injected in-between the two glasses, and measurements initiate with a precise micrometer. Measurements are taken at each 15 s within the first 2 min and at every minute after that [9]. Since, at least during the dehydration, these measurements show poor precision, it is recommended that some sets of measurements should be performed to average final results. To obtain the thickness kinetics for the sample, we must subtract the glasses thickness ($2 \times d_G$) from the measurements made during treatment.

In Chap. 6, how thickness measurements are used to obtain important information relative to the treatments will be explained in detail.

5.2 In Vivo Measurements

To perform in vivo tissue measurements, usually we cannot adopt the transmittance setups described in previous sections. Instead, only reflectance or imaging methods can be applied to avoid invasive procedures. The most common methods used to make in vivo measurements are the diffuse reflectance spectroscopy or imaging. The following subsections will describe these measurement procedures and present some interesting results obtained for in vivo tissues during OC treatments.

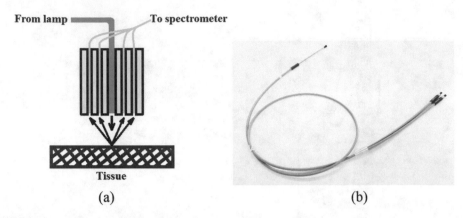

From lamp   **To spectrometer**

Tissue

(a)   (b)

Fig. 5.8 R_d measuring setup: schematics (**a**) and real R_d sensor (**b**)

5.2.1   Diffuse Reflectance

Although total reflectance measurements can be made from in vivo tissues, the most sensitive spectroscopy measurement is the diffuse reflectance (R_d). R_d spectra can be measured during OC treatments, and its kinetics may provide valuable information. To measure R_d spectra from a tissue (ex vivo or in vivo), a setup like the one presented in Fig. 5.8 is used.

From Fig. 5.8b, we see an optical fiber sensor for R_d spectroscopy. The top-left tip of the sensor is the one represented schematically in Fig. 5.8a. Such tip contains the illuminating fiber (at the center in Fig. 5.8a) that brings the light from a broadband lamp. Such beam is incident at the top surface of the sample, and the light that is scattered within a small cone is collected by the other fibers in the tip of the sensor to be delivered to the spectrometer. In Fig. 5.8b we see the two tips in the fiber cable that connect to the lamp and to the spectrometer.

R_d spectra are sensitive to the absorption of tissue components. Figure 5.9 shows a typical R_d spectrum that has been acquired from an ex vivo human colorectal muscle sample.

In Fig. 5.9 we see some absorption bands, which are related to the presence of water and different forms of hemoglobin in the muscle. To identify these absorption bands in the ex vivo muscle, we present in Table 5.1 the absorption bands for the different forms of hemoglobin as indicated in literature and the Internet [35, 36].

All data in Table 5.1 were retrieved from Ref. [36], except the first wavelength for oxyhemoglobin (415 nm), which corresponds to the Soret band and was retrieved from Ref. [35].

Looking into Fig. 5.9, we see that the observable absorption bands for the muscle occur at the central peaks of 415, 493, 543, 620, and 977 nm. Comparing these wavelengths to the ones in Table 5.1, we see that the peaks at 415 and 543 correspond to oxyhemoglobin, the peak at 493 nm is very close to the first absorption peak of

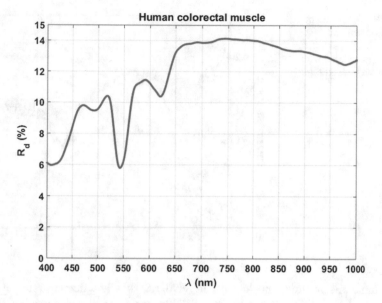

Fig. 5.9 R_d spectrum from a human colorectal muscle sample

Table 5.1 Central absorption peaks for different forms of hemoglobin between 400 and 700 nm [35, 36]

Hemoglobin type	Wavelength (nm)
Oxyhemoglobin	415, 541, 576
Deoxyhemoglobin	431, 556
Carboxyhemoglobin	538, 568
Methemoglobin	500, 630
Sulfhemoglobin	620
Methemalbumin	624
Memochromogen (Schumm test)	558

methemoglobin (500 nm), and the peak at 620 nm corresponds to sulfhemoglobin. The peak at 977 nm corresponds to one of the absorption peaks of water, which usually occurs between 975 and 985, according to data published on Ref. [37]. The muscle sample used to acquire the R_d spectrum in Fig. 5.9 was collected from a surgical specimen of a patient undergoing treatment to remove colorectal cancer at the Oncology Institute of Porto, Portugal. The presence of both methemoglobin and sulfhemoglobin show the inability of hemoglobin to transport oxygen, which may in turn lead to cyanosis [38, 39].

The measurement of R_d spectra from in vivo tissues has also been performed both in natural tissues and undergoing OC treatment. Figure 5.10 shows the R_d spectra from human normal and pathological colon mucosa [40].

The data presented in Fig. 5.10 represents real measured spectra (thicker lines) and calculated fitting model (thin lines). Spectra presented in Fig. 5.10 that correspond both to normal and pathological mucosa show the absorption bands for oxyhemoglobin—415, 542, and 577 nm [40].

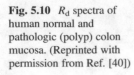

Fig. 5.10 R_d spectra of human normal and pathologic (polyp) colon mucosa. (Reprinted with permission from Ref. [40])

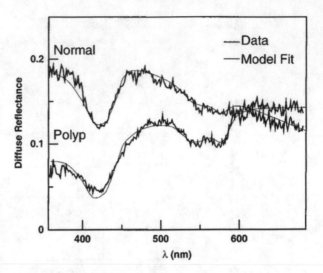

A more recent study made with in vivo rat skin has demonstrated that R_d decreases during OC treatment with 40%-glucose after dermal injection [41]. - Figure 5.11 shows the decrease of the entire R_d spectrum and the R_d kinetics at some wavelengths.

From Fig. 5.11a we see the occurrence of the absorption bands for oxyhemoglobin, which tend to decrease magnitude during the OC treatment. Both graphs in Fig. 5.11 show that R_d decreases during OC treatment, as expected.

We have also observed such decrease in a brief study made with ex vivo human colorectal muscle under treatment with 40%-glycerol—Fig. 5.12.

Figure 5.12 shows that R_d decreases during treatment. Other studies performed from different tissues and under treatment with other OCAs have showed similar behavior. Authors of Ref. [42] have measured R_d spectra from ex vivo rat skin for 20 min after application of anhydrous glycerol to the dermal side and obtained similar results to the ones presented in Figs. 5.11 and 5.12.

Considering measurements made from untreated tissues, Ref. [43] presents a method based on Monte Carlo simulations to estimate the R_d data from normal and pathological tissues in visible-NIR range. By creating a lookup table with such data, the measured R_d spectra are compared to the estimated values so that a diagnosis can be established. Such method was further studied with normal buccal mucosa and oral squamous cell carcinoma tissues, allowing for a 90% sensitivity in discriminating cancer [44].

R_d spectra measured from in vivo skin during OC treatments with aqueous solutions containing different glycerol concentrations have showed time dependence that correlates with sample thickness kinetics obtained from image analysis with software: R_d and skin thickness decrease or increase simultaneously [45].

R_d has also been used to evaluate OC treatments, their mechanisms, and the potential of enhancing agents [46, 47]. Considering human skin treated with different OCAs, R_d at 540 nm presents discriminated decrease at 10 and 60 min—see

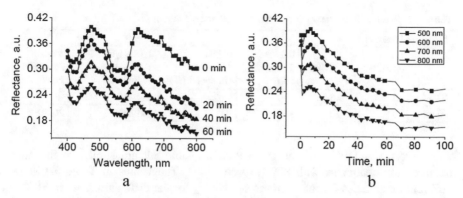

Fig. 5.11 R_d spectra of rat skin in vivo during treatment with 40%-glucose (**a**) and corresponding kinetics for some visible wavelengths (**b**). (Reprinted with permission from Ref. [41])

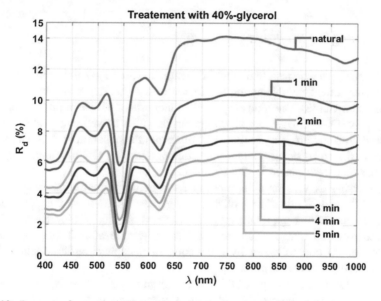

Fig. 5.12 R_d spectra from colorectal muscle during treatment with 40%-glycerol

Table 5.2 [46]. Since Azone is water and oil soluble, a similar study was made to evaluate the enhanced penetration of glycerol with water-Azone (H_2O/Az) and propylene glycol (PG) with oil-Azone (OIL/Az) solutions in porcine skin [47]. The variations observed for R_d after 60 min are presented in the last two columns of Table 5.2.

The enhancement indicated in the last column of Table 5.2 is produced by the application of sodium lauryl sulfate (common pharmaceutical and cosmetics penetration enhancer) and application of ultrasound [46]. For the enhanced treatments

Table 5.2 Reduction in skin R_d at 540 nm after 10 and 60 min of OC [46]

Time	Treatment					
	No treatment (%)	PEG400 (%)	0.25%-Thiazone (%)	0.25%-Thiazone/ enhanced (%)	5%-H$_2$O/ Az + 40%- Glycerol	5%-oil/ Az + 40%- PG
10 min	~0	~5	~7	~16	–	–
60 min	~0	~20	~35	~41	29.3%	20.6%

with glycerol and PG, the variations seen in Table 5.2 are similar to the ones obtained for treatments with 80% glycerol [47]. The values in Table 5.2 show different magnitudes for the decrease of skin R_d for the different OCAs used. It is clear from these data that when OCA penetration is enhanced, variations present higher magnitude.

5.2.2 Imaging Methods

Imaging methods when applied with OC treatments are in general used to study the transparency effect, by evaluating the increase in tissue depth and contrast of the images obtained [48]. One study reported an increase of 41.3% in optical coherence tomography (OCT) imaging depth on human skin after treated with 0.25%-thiazone enhanced with sodium lauryl sulfate and ultrasound application [46]. Such methods can be used also to obtain certain measurements during OC treatments or global variations that were caused by the treatments.

Considering ex vivo rat skin, a comparison between images taken from native sample, 1, 5, and 20 min after application of glycerol, shows that the skin increases its transparency along the treatment [42]. The reversibility of the treatment is also observed from an image taken from the same sample after it was treated with glycerol for 20 min followed by immersion in phosphate-buffered saline (PBS) solution during another 20 min.

Another OCT study evaluated thickness change in gastric tissues under treatment with 50% glycerol solutions [49]. Sections of tissue were placed on a glass slide for OCT image acquisition. The sample thickness from each image was calculated as the difference between the axial positions of the top of the glass slide and the axial position of sample surface. Images were taken before and after 50%-glycerol topical application at the time interval of 20, 40, and 60 min. After averaging results from four independent studies, sample thickness showed a decrease of about 20%, which was more significant within the first 20 min of treatment [49]. Figure 2 of Ref. [46] shows transmission electron microscopy images where the ultrastructure details of tendon fibrils are visible for native sample and after glycerol treatment or exposed to the air. The air-exposed sample was only dehydrated, but the fibril packing is similar for glycerol and air-treated samples, showing higher packing density than the native sample [46].

Although the measurements presented in the above cases were performed from ex vivo samples, imaging methods can be easily applied to perform similar measurements from in vivo tissues.

References

1. B. Chance, M. Cope, E. Gratton, N. Ramanujam, B. Tromberg, Phase measurement of light absorption and scatterer in human tissue. Rev. Sci. Instrum. **69**(10), 3457–3481 (1998)
2. A.V. Priezzhev, V.V. Tuchin, L.P. Shubochkin, *Laser Diagnostics in Biology and Medicine* (Nauka, Moscow, 1989)
3. V.V. Tuchin, *Lasers and Fiber Optics in Biomedical Science*, 2nd edn. (Saratov University Press, Saratov, 2010)
4. V.V. Tuchin, *Optical Clearing of Tissues and Blood* (SPIE Press, Bellingham, 2006)
5. V.V. Tuchin, *Tissue Optics: Light Scattering Methods and Instruments for Medical Diagnosis*, 3rd edn. (SPIE Press, Bellingham, 2015)
6. L. Oliveira, M.I. Carvalho, E.M. Nogueira, V.V. Tuchin, Optical clearing mechanisms characterization in muscle. J. Innov. Opt. Health Sci. **9**(5), 1650035 (2016)
7. I. Carneiro, S. Carvalho, R. Henrique, L. Oliveira, V.V. Tuchin, et al., IEEE J. Sel. Top. Quant. Electron. **25**(1), 7200608 (2019)
8. I. Carneiro, S. Carvalho, R. Henrique, L. Oliveira, V.V. Tuchin, A robust ex vivo method to evaluate the diffusion properties of agents in biological tissues. J. Biophotonics **12**(4), e201800333 (2019)
9. S. Carvalho, N. Gueiral, E. Nogueira, R. Henrique, L. Oliveira, V.V. Tuchin, Glucose diffusion in colorectal mucosa – a comparative study between normal and cancer tissues. J. Biomed. Opt. **22**(9), 091506 (2017)
10. A.N. Bashkatov, K.V. Berezin, K.N. Dvoretskiy, M.L. Chernavina, E.A. Genina, V.D. Genin, V.I. Kochubey, E.N. Lazareva, A.B. Pravdin, M.E. Shvachkina, P.A. Timoshina, D.K. Tuchina, D.D. Yakovlev, D.A. Yakovlev, I.Y. Yanina, O.S. Zhernovaya, V.V. Tuchin, Measurements of tissue optical properties in the context of tissue optical clearing. J. Biomed. Opt. **23**(9), 091416 (2018)
11. L. Oliveira, M.I. Carvalho, E. Nogueira, V.V. Tuchin, Skeletal muscle dispersion (400–1000 nm) and kinetics at optical clearing. J. Biophotonics **11**(1), e201700094 (2018)
12. I. Carneiro, S. Carvalho, V. Silva, R. Henrique, L. Oliveira, V.V. Tuchin, Kinetics of optical properties of human colorectal tissues during optical clearing: a comparative study between normal and pathological tissues. J. Biomed. Opt. **23**(12), 121620 (2018)
13. S. Carvalho, N. Gueiral, E. Nogueira, R. Henrique, L. Oliveira, V.V. Tuchin, Comparative study of the optical properties of colon mucosa and colon precancerous polyps between 400 and 1000 nm, in *Dynamics and Fluctuations in Biomedical Photonics XIV*, Proc. of SPIE, ed. by V. V. Tuchin, K. V. Larin, M. J. Leahy, R. K. Wang, vol. 10063, (SPIE Press, Bellingham, 2017), p. 100631L. https://doi.org/10.1117/12.2253023
14. I. Carneiro, S. Carvalho, R. Henrique, L. Oliveira, V.V. Tuchin, Optical properties of colorectal muscle in visible/NIR range, in *Biophotonics: Photonic Solutions for Better Health Care VI*, Proc. of SPIE, ed. by J. Popp, V. V. Tuchin, F. S. Pavone, vol. 10685, (SPIE Press, Bellingham, 2018), p. 106853D. https://doi.org/10.1117/12.2306586
15. I. Carneiro, S. Carvalho, R. Henrique, L. Oliveira, V.V. Tuchin, Measuring optical properties of human liver between 400 and 1000 nm. Quant. Electron. **49**(1), 13–19 (2019)
16. A.N. Bashkatov, E.A. Genina, M.D. Kosintseva, V.I. Koshubey, S.Y. Gorodkov, V.V. Tuchin, Optical properties of peritoneal biological tissues in the range of 350–2500 nm. Opt. Spectrosc. **120**(1), 1–8 (2016)

17. A.N. Bashkatov, E.A. Genina, V.I. Koshubey, V.V. Tuchin, Optical properties of human skin, subcutaneous and mucous tissues in the wavelength range from 400 to 2000 nm. J. Phys. D Appl. Phys. **38**(15), 2543–2555 (2005)

18. A.N. Bashkatov, E.A. Genina, V.I. Kochubey, E.A. Kolesnikova, V.V. Tuchin, Optical properties of human colon tissues in the 350–2500 spectral range. Quant. Electron. **44**(8), 779–784 (2014)

19. L.-H. Wang, S.L. Jacques, L.-Q. Zheng, MCML – Monte Carlo modeling of photon transport in multi-layered tissues. Comput. Met. Progr. Biomed. **47**(2), 131–146 (1995)

20. S.A. Prahl, M.J.C. Van Gemert, A.J. Welch, Determining the optical properties of turbid media by using the adding-doubling method. Appl. Opt. **32**(4), 559–568 (1993)

21. S. Peña-Llopis, J. Brugarolas, Simultaneous isolation of high-quality DNA, RNA, miRNA and proteins from tissues for genomic applications. Nat. Protoc. **8**(11), 2240–2255 (2013)

22. R.R. Anderson, J.A. Parish, Optical properties of human skin, in *The Science of Photomedicine*, ed. by J. D. Regan, J. A. Parish, (Plenum Press, New York, 1982), pp. 147–194

23. S. Carvalho, N. Gueiral, E. Nogueira, R. Henrique, L. Oliveira, V.V. Tuchin, Wavelength dependence of the refractive index of human colorectal tissues: comparison between healthy mucosa and cancer. J. Biomed. Photon. Eng. **2**(4), 040307 (2016)

24. I. Carneiro, S. Carvalho, R. Henrique, L. Oliveira, V.V. Tuchin, Water content and scatterers dispersion evaluation in colorectal tissues. J. Biomed. Photon. Eng. **3**(4), 040301 (2017)

25. H. Ding, J.Q. Lu, W.A. Wooden, P.J. Kragel, X.H. Hu, Refractive indices of human skin tissues at eight wavelengths and estimated dispersion relations between 300 and 1600 nm. Phys. Med. Biol. **51**(6), 1479–1489 (2006)

26. Z. Deng, J. Wang, Q. Ye, T. Sun, W. Zhou, J. Mei, C. Zhang, J. Tian, Determination of continuous complex refractive index dispersion of biotissue based on internal reflection. J. Biomed. Opt. **21**(1), 015003 (2016)

27. S. Liu, Z. Deng, J. Li, J. Wang, N. Huang, R. Cui, Q. Zhang, J. Mei, W. Zhou, C. Zhang, Q. Ye, J. Tian, Measurement of the refractive index of whole blood and its components for a continuous spectral region. J. Biomed. Opt. **24**(3), 035003 (2019)

28. X.U. Zhang, D.J. Faber, A.L. Post, T.G. van Leeuwen, H.J.C.M. Sterenborg, Refractive index measurement using single fiber reflectance spectroscopy. J. Biophoton. **12**(7), e201900019 (2019)

29. G. Vargas, K.F. Chan, S.L. Thomsen, A.J. Welch, Use of osmotically active agents to alter optical properties of tissue: effects on the detected fluorescence signal measured through skin. Laser Surg. Med. **29**, 213–220 (2001)

30. E.A. Genina, A.N. Bashkatov, A.A. Zubkova, V.V. Tuchin, I. Yaroslavsky, G.B. Altshuler, Optical clearing of human skin: comparative study of permeability and dehydration of intact and photothermally perforated skin. J. Biomed. Opt. **13**(2), 021102 (2008)

31. D.K. Tuchina, A.N. Bashkatov, A.B. Bucharskaya, E.A. Genina, V.V. Tuchin, Study of glycerol diffusion in skin and myocardium ex vivo under the conditions of developing alloxan-induced diabetes. J. Biomed. Photon. Eng. **3**(2), 020302 (2017)

32. V.D. Genin, D.K. Tuchina, A.N. Bashkatov, E.A. Genina, V.V. Tuchin, Polyethylene glycol diffusion in ex vivo skin tissue. AIP Conf. Proc. **1688**, 030028 (2015)

33. L.M. Oliveira, M.I. Carvalho, E.M. Nogueira, V.V. Tuchin, The characteristic time of glucose diffusion measured for muscle tissue at optical clearing. Laser Phys. **23**, 075606 (2013)

34. L.M. Oliveira, M.I. Carvalho, E.M. Nogueira, V.V. Tuchin, Diffusion characteristics of ethylene glycol in skeletal muscle. J. Biomed. Opt. **20**(5), 051019 (2015)

35. https://omlc.org/spectra/hemoglobin/. Accessed 8 July 2019

36. L.D. Robertson, D. Roper, Laboratory methods used in the investigation of the haemolytic anaemias, in *Dacie and Lewis Practical Haematology*, ed. by B. J. Bain, I. Bates, M. A. Laffan, 11th edn., (Elsevier, Amsterdam, 2011)

37. G.M. Hale, M.R. Querry, Optical constants of water in the 200nm to 200 micron wavelength region. Appl. Opt. **12**, 555–563 (1973)

38. https://en.wikipedia.org/wiki/Methemoglobin. Accessed 8 July 2019

39. https://en.wikipedia.org/wiki/Sulfhemoglobinemia. Accessed 8 July 2019
40. G. Zonios, L.T. Perelman, V. Backman, R. Mahoharan, M. Fitzmaurice, J. van Dam, M.S. Feld, Diffuse reflectance spectroscopy of human adenomatous colon polyps in vivo. Appl. Opt. **38** (31), 6628–6637 (1999)
41. D.K. Tuchina, P.A. Timoshina, V.V. Tuchin, A.N. Bashkatov, E.A. Genina, Kinetics of rat skin optical clearing at topical application of 40%Glucose: ex vivo and in vivo studies. IEEE J. Sel. Top. Quant. Electron. **25**(1), 7200508 (2019)
42. G. Vargas, E.K. Chan, J.K. Barton, H.G. Rylander, A.J. Welch, Use of an agent to reduce scattering in skin. Laser Surg. Med. **24**, 133–141 (1999)
43. G. Einstein, P. Aruna, S. Ganesan, Monte Carlo based model for diffuse reflectance from turbid media for the diagnosis of epithelial dysplasia. Optik **181**, 828–835 (2018)
44. G. Einstein, K. Udayakumar, P.R. Aruna, D. Koteeswaran, S. Ganesan, Diffuse reflectance spectroscopy for monitoring physiological and morphological changes in oral cancer. Optik **127**, 1479–1485 (2016)
45. X. Wen, Z. Mao, Z. Han, V.V. Tuchin, D. Zhu, In vivo skin optical clearing by glycerol solutions: mechanism. J. Biophotonics **3**(1–2), 44–52 (2010)
46. D. Zhu, K.V. Larin, Q. Luo, V.V. Tuchin, Recent progress in tissue optical clearing. Laser Photon. Rev. **7**(5), 732–757 (2013)
47. X. Xu, Q. Zhu, Evaluation of skin optical clearing enhancement with Azone as a penetration enhancer. Opt. Commum. **279**, 223–228 (2007)
48. A. Sdobnov, M.E. Darvin, E.A. Genina, A.N. Bashkatov, J. Lademan, V.V. Tuchin, Recent progress in tissue optical clearing for spectroscopic application. Spectrochim. Acta A Mol. Biomol. Spectrosc. **197**, 216–229 (2018)
49. X. Xu, R. Wang, J.B. Elder, Optical clearing effect on gastric tissues immersed with biocompatible chemical agents investigated by near infrared reflectance spectroscopy. J. Phys. D Appl. Phys. **36**, 1707–1713 (2003)

Chapter 6
Data that Can Be Acquired from Optical Clearing Studies

6.1 Introduction

In the previous chapter, we have discussed several types of measurements that can be acquired during optical clearing (OC) treatments from ex vivo or in vivo tissues. Such measurements can be used to calculate various forms of data that have interest to characterize the tissue or the treatment. Regarding the tissue, information about the mobile water content may become available, and pathology identification is possible.

In the following sections, we will refer to the measurements described in the previous chapter to explain how they can be used to obtain such data from OC treatments.

6.2 Evaluation of the OC Mechanisms Through the Refractive Index Kinetics

According to the description presented in Chap. 4 for the OC mechanisms, we know that any treatment is driven by the water flux out and the optical clearing agent (OCA) flux into the tissue. These two fluxes originate the partial exchange of the interstitial water by the OCA, which creates an increase in the refractive index (RI) of the background material of the tissue—the interstitial fluid (ISF) [1–7].

An evaluation of the kinetics of the two fluxes and also of the increase in the RI of the ISF to produce the RI matching mechanism can be obtained from collimated transmittance (T_c) and thickness measurements made during the treatment [1, 8, 9]. These measurements are sensitive to the water flux out and to the OCA flux in, meaning that measurements performed during an OC treatment can be used to evaluate the RI variations. To explain the sequence of calculations to quantify the

© The Author(s), under exclusive license to Springer Nature Switzerland AG 2019
L. M. C. Oliveira, V. V. Tuchin, *The Optical Clearing Method*,
SpringerBriefs in Physics, https://doi.org/10.1007/978-3-030-33055-2_6

Fig. 6.1 Histology image
of the human colorectal wall
showing the various
layers—mucosa,
submucosa, and *muscularis
propria* (photo taken at the
Portuguese Oncology
Institute of Porto by our
colleague Dr. Sónia
Carvalho)

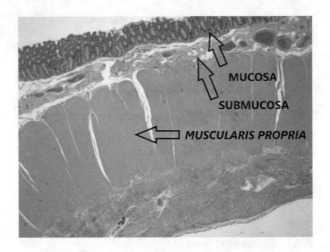

RI kinetics for the ISF and for the whole tissue, an experimental example will be presented.

In this study, samples from the muscle layer of human colorectal wall are used. Figure 6.1 shows a histological image of the colorectal wall where all layers are seen.

The samples used in this study were obtained from the colorectal resections of four patients under treatment at the Portuguese Oncology Institute of Porto, Portugal. The patients have previously signed a written consent to use surgical specimens for diagnostic and research purposes. The muscle layer (*muscularis propria* in Fig. 6.1) was retrieved from the surgical resections, and using a cryostat (Leica model CM 1850 UV), the muscle layer was fragmented into six smaller samples with slab form ($\phi = 1$ cm and $d = 5$ mm) [8].

Three of these samples were used to measure thickness kinetics during treatment with 40%-glycerol. Those measurements were made according to the procedure described in Sect. 5.1.5. The other three samples were used to measure collimated transmittance (T_c) kinetics during treatment with the same solution. The procedure adopted in these measurements is described in Sect. 5.1.1, and the average results obtained in both types of measurements are presented in Fig. 6.2.

After completing the measurements, the average results in Fig. 6.2 were used in calculations. Considering first the native muscle, we started by calculating the volume fractions (VFs) for the scatterers and the ISF. According to our model, scatterers in muscle tissue are a combination of dried muscle fibers and water that is bound to the fibers to keep them hydrated. This water is designated as bound water. The ISF considered in our model is composed only by water that can move during the treatments, which is designated as mobile water [8, 9]. We also considered that for short-term OC treatments, the bound water is not converted into mobile water. This means that the absolute volume of scatterers and their RI is kept unchanged during the treatment. It is important to know how much water in a tissue is able to move during treatments, and some recent studies have been made to understand the various states of water in a biological tissue. These studies have reported that

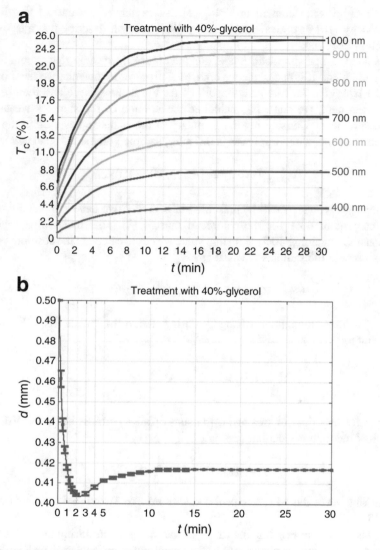

Fig. 6.2 T_c (**a**) and thickness (**b**) kinetics of the human colorectal muscle tissue under treatment with 40%-glycerol. (Reprinted with permission from Ref. [9])

tissues are known to have four states of water, depending on their bounding strength: strongly bound, tightly bound, weakly bound, and free water [6, 10]. Only the last two states are considered as mobile water, and they participate in the dehydration mechanism during short-term OC treatments, meaning that the other two states remain bound to the other tissue components [10, 11]. Mobile water can be found both in the ISF and inside tissue cells, and to move it through cell membranes into the ISF, a strong or long-term osmotic pressure is necessary [11].

The mobile water content in biological tissues can be evaluated using Raman spectroscopy [10] or optical spectroscopy [11]. We will describe the later method in detail in Sect. 6.4.

Considering our simplistic model for the colorectal muscle that accounts only for scatterers and ISF and since we applied short-term treatments, we need only to consider two VFs. To calculate those VFs for the natural tissue and their kinetics during treatment, we started to calculate the volume of the samples we used in the measurements. Considering the geometry of the slab samples used in the study (diameter is 1.0 cm, and thickness is 0.5 mm), sample volume is calculated as:

$$V_{sample}(t = 0) = \left(\pi \times 0.5^2\right) \times 0.05\,\mathrm{cm}^3. \tag{6.1}$$

From a study, which will be presented in Sect. 6.4, we have estimated a mobile water content of 60% for the colorectal muscle [8]. This means that scatterers represent the remaining 40%. With this value we could calculate the absolute volume for scatterers in the samples we used as:

$$V_s = 0.4 \times V_{sample}(t = 0)\,\mathrm{cm}^3. \tag{6.2}$$

This volume remains unchanged during treatment, while its VF changes according to:

$$f_s(t) = \frac{V_s}{\left(\pi \times 0.5^2\right) \times d(t)}. \tag{6.3}$$

Since for any time of treatment, the sum of all VFs must be 1; the VF of the interstitial medium is calculated as:

$$f_{ISF}(t) = 1 - f_s(t). \tag{6.4}$$

The kinetics of the VFs, as calculated with Eqs. (6.3) and (6.4), are presented in Fig. 6.3.

As it is seen from Fig. 6.3, the strongest variations occur during the first 2 min of treatment, indicating the occurrence of the dehydration mechanism. The RI matching mechanism produces smoother and small magnitude variations in the VFs.

After calculating the kinetics for the VFs of tissue components, the kinetics for the scattering coefficient (μ_s) of the human colorectal muscle was calculated. In this calculation, the average $T_c(t)$ and $d(t)$ data presented in Fig. 6.2 were used. The estimated data for the absorption coefficient ($\mu_a(\lambda)$) of the natural tissue obtained in previous study [12] were also used. The calculation was made using Beer-Lambert equation [8, 11]:

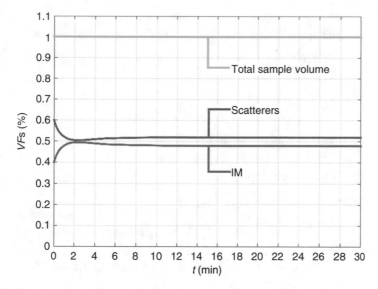

Fig. 6.3 Kinetics of the VFs for the human colorectal muscle tissue during treatment with 40%-glycerol. (Reprinted with permission from Ref. [9])

$$\mu_s(\lambda, t) = -\frac{\ln\left[T_c(\lambda, t)\right]}{d(t)} - \mu_a(\lambda). \tag{6.5}$$

Since in biological tissues μ_s is in general much higher than μ_a [12–15], it is reasonable to consider in calculations made with Eq. (6.5) that μ_a remains unchanged. Figure 6.4 presents the kinetics for the attenuation coefficient ($\mu_t = \mu_a + \mu_s$) and μ_s at 650 nm.

According to Fig. 6.4, the variations that occur during treatment for μ_t and μ_s are high. For $t = 0$ (native tissue) we observe a small difference between these properties, meaning that μ_a is small when compared to μ_s, and any variations in μ_a can be neglected in this calculation. An increase at the beginning of the treatment (first 15 s) is seen, which indicates that due to water loss, scatterers approach each other (providing tissue shrinkage), leading to the increase of both coefficients but to the decrease of tissue sample thickness.

The following step is to calculate the variations in the RI of the ISF. According to literature [2, 8, 9], for a particular wavelength, the RI of the ISF ($\bar{n}_0$) can be calculated for any time of treatment (t) according to Eq. (6.6):

$$\bar{n}_0(t) = \frac{n_s}{\left(\sqrt{\frac{\mu_s(t) \times d(t)}{\mu_s(t=0) \times d(t=0)}} \times \left(\frac{n_s}{n_0(t=0)} - 1\right) + 1\right)}, \tag{6.6}$$

where, for that particular wavelength, n_s is the RI of tissue scatterers, $\mu_s(t)$ represents the scattering coefficient of the tissue at time t, and $d(t)$ represents the sample thickness at time t. For the natural tissue ($t = 0$), the scattering coefficient is

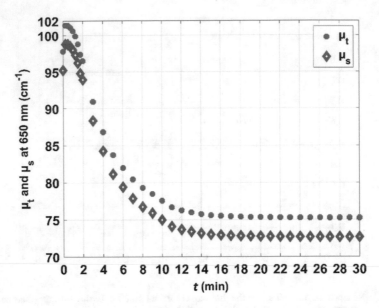

Fig. 6.4 Kinetics of μ_s and μ_t at 650 nm for the human colorectal muscle tissue during treatment with 40%-glycerol. (Reprinted with permission from Ref. [9])

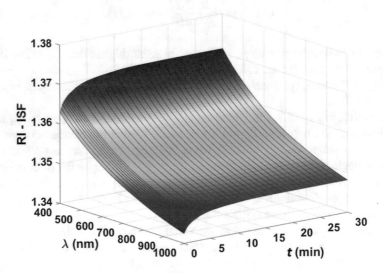

Fig. 6.5 Kinetics of the RI of the ISF for the human colorectal muscle tissue during treatment with 40%-glycerol. (Reprinted with permission from Ref. [9])

represented by $\mu_s(t = 0)$, and the sample thickness is represented by $d(t = 0)$ [8]. This calculation was made for several wavelengths between 400 and 1000 nm, and the results are presented in Fig. 6.5.

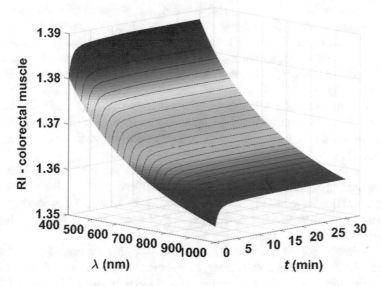

Fig. 6.6 Kinetics of the RI for the whole human colorectal muscle tissue during treatment with 40%-glycerol. (Reprinted with permission from Ref. [9])

As a result of the treatment, Fig. 6.5 shows a smooth increase of the RI of ISF for all wavelengths, and it proves the occurrence of the RI matching mechanism.

At this stage, one is able to use the Gladstone and Dale law of mixtures [2, 16–19] to calculate the kinetics of the RI of the whole tissue ($n_{\text{tissue}}(\lambda, t)$) during treatment:

$$n_{\text{tissue}}(\lambda, t) = \bar{n}_0(\lambda, t) f_{\text{ISF}}(t) + n_s(\lambda) f_s(t). \qquad (6.7)$$

The data calculated with Eq. (6.7) is presented in Fig. 6.6.

Figure 6.6 shows similar increasing behavior to the one observed in Fig. 6.5 but with different magnitude in variations. In this figure, we see three stages: within the first minute, a strong increase in n_{tissue} is created by the dehydration mechanism. The following 3 min correspond to a transition between mechanisms, and final 26 min correspond uniquely to the RI matching mechanism. Figure 6.6 also shows that since glycerol has no strong absorption bands in this wavelength range, it produces similar efficiency in the entire spectral range.

6.3 Estimation of the Kinetics for Tissue's Optical Properties During OC Treatments

The estimation of the kinetics of the optical properties of a tissue under OC treatment is also a way to characterize the treatment efficiency. The increase in tissue transparency is created through the reduction of the scattering coefficient and increase of the anisotropy factor [2].

One way to obtain such kinetics relies on performing estimations with inverse Monte Carlo [20] or inverse adding-doubling [21] software for various times of treatment and for various wavelengths. Such large amount of simulations is time-consuming and involves the need to perform various sets of measurements during treatment, namely, collimated transmittance (T_c), total transmittance (T_t), and total reflectance (R_t) [11, 14].

A much simpler method, based on Mie scattering theory [22, 23], can be used. This calculation method is based on spherical scatterers, but it can approximate real tissue scatterers with cylindrical shape where the cylinders are not very elongated [11].

Considering the example for colorectal muscle under treatment with 40%-glycerol and the calculations presented in the previous section, we already have the kinetics for μ_s. In Fig. 6.4, we presented the kinetics of μ_s at 650 nm, but considering the entire spectral range used in that study, the kinetics of μ_s between 400 and 1000 nm can be presented.

Again, in Fig. 6.7 we see a small increase in μ_s within the first 15 s of treatment as a result of scatterers approaching each other. After this initial increase, we see smooth decrease at all wavelengths. The 3D map presented in Fig. 6.7 shows smooth wavelength dependence at any time of treatment.

In the previous section, we calculated also the kinetics for the RI of the ISF ($\bar{n}_0$) and considered that the wavelength dependence for the RI of tissue scatterers ($n_s(\lambda)$) keeps unchanged during treatment. These data can be used to calculate the kinetics of the relative RI of the muscle tissue. This parameter is commonly used to characterize the scattering efficiency of biological tissues and is defined according to Eq. (6.8) [2, 24]:

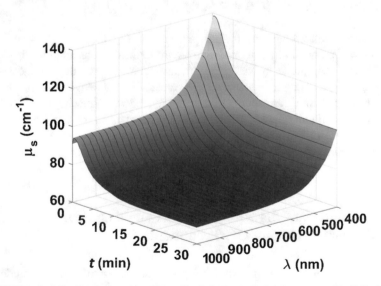

Fig. 6.7 Kinetics of μ_s for the human colorectal muscle tissue during treatment with 40%-glycerol. (Reprinted with permission from Ref. [9])

$$m = \frac{n_s}{\bar{n}_0}. \tag{6.8}$$

High and multiple light scattering occurs in most biological tissues, since they have $m > 1$ [2]. For a tissue to become completely transparent, m should decrease to 1, and the reduced scattering coefficient (μ_s') should decrease to zero.

According to the Mie scattering theory, μ_s' can be defined as a function of m, $\bar{n}_0$, light wavelength (λ), the mean scatterer radius (a), and the volume density of the scattering centers (ρ_s) [22, 23]. Equation (1.18) establishes such relation, and we reproduce it here due to its major importance in the evaluation of the kinetics for μ_s' during OC treatments:

$$\mu_s' = 3.28\pi a^2 \rho_s \left(\frac{2\pi\bar{n}_0 a}{\lambda}\right)^{0.37} (m-1)^{2.09}. \tag{6.9}$$

As already indicated in Chap. 1, Eq. (6.9) was validated by the authors of Ref. [22] for the visible-near-infrared (NIR) spectral range and for a system of noninteracting spherical particles with a mean diameter of $2a$ and when scattering anisotropy factor $g > 0.9$, $5 < 2\pi a/\lambda < 50$, and $1 < m < 1.1$. Recently, this equation was tested outside those limits to estimate the kinetics of μ_s' for normal and pathological human colorectal mucosa, and the calculated results are accordingly to what is predicted in literature [11]. Also for the human colorectal muscle tissue, the estimated kinetics shows the expected behavior.

Figure 6.8 demonstrates the kinetics for m, as calculated by Eq. (6.8).

As we see from Fig. 6.8, relative RI m not only decreases with the wavelength but also decreases during the time of treatment in a smooth fashion. Such time

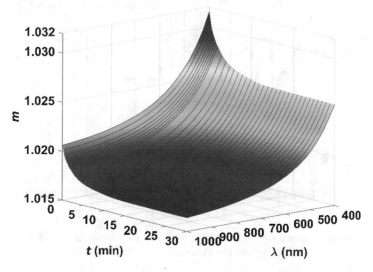

Fig. 6.8 Kinetics of the relative RI m for the human colorectal muscle tissue during treatment with 40%-glycerol. (Reprinted with permission from Ref. [9])

dependence shows that due to the increase in $\bar{n}_0$, scattering reduces in the tissue for all wavelengths, as we have already seen in Fig. 6.7.

To proceed with calculation of the kinetics for μ'_s, one needs first to determine the value for a in the case of colorectal muscle tissue. To determine such value, we rearranged part of Eq. (6.9) in a way that ρ_s is expressed as a function of the VF of tissue scatterers.

$$\mu'_s = \frac{3f_s(1-f_s)}{4\pi a^3} \times 3.28\pi a^2 \left(\frac{2\pi\bar{n}_0 a}{\lambda}\right)^{0.37} (m-1)^{2.09}. \qquad (6.10)$$

Equation (6.10) was then used to reconstruct the μ'_s data for the natural tissue that was previously estimated through IAD simulations [12]. In this reconstruction, a good data fitting with $a = 500$ nm was obtained [9]. The estimated data and the reconstructed curve for $\mu'_s(\lambda)$ of the natural (none treated) tissue are presented in Fig. 6.9.

Figure 6.9 shows a very good fit of the estimated data with the curve calculated with Eq. (6.10). Such good reconstruction of $\mu'_s(\lambda)$ data was obtained from the dispersions of scatterers and ISF. Since a model that considers unchanged volume for tissue scatterers was assumed, the estimated a value remains unchanged during treatment.

To calculate $\mu'_s(\lambda)$ data for various times of treatment, we used (Eq. 6.10) again, but now replacing f_s by $f_s(t)$, and $\bar{n}_0$ and m by the data from Figs. 6.5 and 6.8. The time dependence for $\mu'_s(\lambda)$ that was obtained in those calculations is presented in Fig. 6.10.

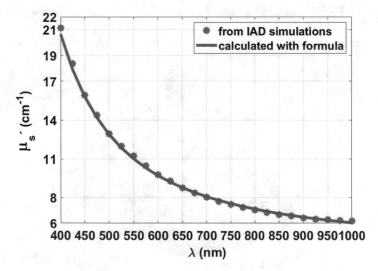

Fig. 6.9 Reduced scattering coefficient $\mu'_s(\lambda)$ data for the human colorectal muscle tissue: estimated by IAD simulations (points) and calculated by Eq. (6.10) (curve)

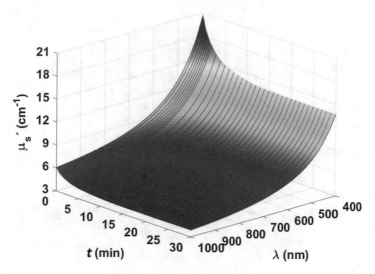

Fig. 6.10 Kinetics of $\mu_s'(\lambda)$ data for the human colorectal muscle tissue during treatment with 40%-glycerol. (Reprinted with permission from Ref. [9])

The calculated $\mu_s'(\lambda, t)$ data present similar behavior to the one obtained for m (see Fig. 6.8). Analyzing Fig. 6.10, we see that the calculations with Eq. (6.10) generate data that decreases during treatment as expected and accordingly to what was already obtained for μ_s (see Fig. 6.7).

Finally, using the data from Figs. 6.7 and 6.10, the kinetics of the anisotropy factor, $g(\lambda, t)$ can be calculated. To perform this calculation, the following relation between these optical properties was used [9, 11]:

$$g = 1 - \frac{\mu_s'}{\mu_s}.\qquad(6.11)$$

The calculated 3D map for g is presented in Fig. 6.11.

We can see from Fig. 6.11 that g-factor increases strongly during the first 2 min of treatment and then it decreases a little before stabilizing for the rest of the treatment. Such behavior can be explained as the following. At early stage of treatment, as the tissue dehydrates, scatterers approach each other and rearrange in an ordered form, increasing g-factor. As the glycerol molecules diffused into tissue sample start to force the scatterers to separate again, some of the scatterers may change a little their orientation, leading to the small decrease observed for g-factor after 2 min of treatment. At later stage of treatment, when the RI matching is the only mechanism taking action inside, g-factor tends to stabilize.

Considering the kinetics obtained in this example for the fundamental optical properties, additional calculations could be made to derive the kinetics of other optical properties such as the light penetration depth. On the other hand, similar calculations can be made for other tissues under treatment with other solutions. The

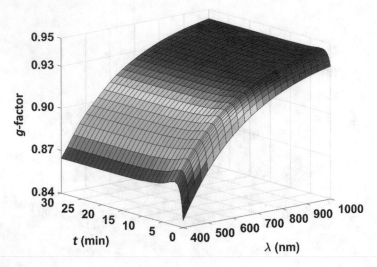

Fig. 6.11 Kinetics of the *g*-factor of the human colorectal muscle tissue during treatment with 40%-glycerol. (Reprinted with permission from Ref. [9])

evaluation of such kinetics provides information on the magnitude of the changes created in the optical properties for a particular treatment and may be useful to plan some OC treatments to be applied in clinical procedures.

6.4   Estimation of the Diffusion Properties of Water and Agents in Tissues

The characterization of chemicals, drugs, oils, and lotions into biological materials is of great interest for many fields of biomedical engineering, such as food industry, organ cryopreservation, clinical practice, pharmacology, or cosmetics [7].

In the field of Biophotonics, and in particular, in the OC method, the evaluation of the diffusion properties of OCAs and water in different tissues is also necessary [7, 24]. If for a particular treatment we are able to evaluate the diffusion time (τ) and the diffusion coefficient (D), we can characterize the water and OCA fluxes with these properties. Consequently, since the tissue dehydration and the RI matching mechanisms are directly associated with the water flux out and the OCA flux into the tissue, these mechanisms are also characterized through these properties.

Some studies have been performed with optical coherence tomography (OCT) to obtain the diffusion properties of fluids in biological tissues [25–29]. Using this method, authors of Ref. [25] have reported glucose's permeability rate ($P = D/l$, where l is the membrane thickness) in in vitro rabbit sclera for upper 80–100 μm and for deeper 100 μm regions: $(6.01 \pm 0.37) \times 10^{-6}$ cm/s and $(2.84 \pm 0.68) \times 10^{-5}$ cm/s, respectively. The authors of Ref. [26] used the OCT technique to evaluate glucose diffusivity in human normal and cancerous esophagus tissue ex vivo: $(1.74 \pm 0.04) \times 10^{-5}$ cm/s and $(2.45 \pm 0.06) \times 10^{-5}$ cm/s, respectively.

If ex vivo tissues are available to evaluate the diffusivity of agents, a simple method can be used [7, 30–34]. To explain this method, we consider a slab-form tissue sample with thickness d, under treatment with an aqueous solution containing an OCA, whose molecules can flow through both surfaces into the tissue in a free diffusion regime. Such diffusion can be described by Fick's law [2, 24]:

$$\frac{\partial C_{OCA}(x,t)}{\partial t} = D_{OCA} \frac{\partial^2 C_{OCA}(x,t)}{\partial x^2}. \tag{6.12}$$

In Eq. (6.12), $C_{OCA}(x,t)$ represents the OCA concentration inside the tissue, D_{OCA} is the OCA diffusion coefficient, x is the unidirectional position between the two slab surfaces, and t represents time. The relation between D and τ can be written for the cases where the diffusion occurs through both slab surfaces (Eq. 6.13) or only through one slab surface (Eq. 6.14) [2].

$$\tau_{OCA} = \frac{d^2}{\pi^2 D_{OCA}}, \tag{6.13}$$

$$\tau_{OCA} = \frac{4d^2}{\pi^2 D_{OCA}}. \tag{6.14}$$

When the volume of the solution is significantly higher than the volume of the slab tissue (e.g., 10×), we can determine the amount of dissolved matter m_t at an instant t relative to its equilibrium value m_∞ according to the following definition [2]:

$$\frac{m_t}{m_\infty} = \frac{\int_0^d C_{OCA}(x,t)dx}{C_{OCA0} \times d}$$

$$= 1 - \frac{8}{\pi^2} \left[\exp\left(-\frac{t}{\tau_{OCA}}\right) + \frac{1}{9} \exp\left(-\frac{9t}{\tau_{OCA}}\right) \right.$$

$$\left. + \frac{1}{25} \exp\left(-\frac{25t}{\tau_{OCA}}\right) + \cdots \right] \tag{6.15}$$

The ratio in Eq. (6.15) represents the volume averaged concentration of an OCA $C_{OCA}(x,t)$ within the slab at time t. Considering a first-order approximation, Eq. (6.15) has a solution given by [2, 35]:

$$C_{OCA}(t) = \frac{1}{d} \int_0^d C_{OCA}(x,t)dx \cong C_{OCA0} \left[1 - \exp\left(-\frac{t}{\tau_{OCA}}\right) \right]. \tag{6.16}$$

Equation (6.16) shows a relation between the time dependence of OCA concentration in the tissue and the characteristic diffusion time τ_{OCA}. If one has means to determine τ_{OCA}, it can be used in Eq. (6.13) or (6.14) to obtain the characteristic D_{OCA}.

If one measures collimated transmittance T_c spectra from a slab tissue under immersion in a solution containing an OCA, the time dependence for T_c at some wavelengths can be calculated. Such curves will have the behavior represented in Fig. 6.2a. By displacing these curves to have $T_c = 0$ at $t = 0$ and then normalizing the data in each curve to its highest value, one can describe the displaced and normalized data with Eq. (6.17) [7, 30, 31]:

$$T_c(\lambda, t) \cong \left[1 - \exp\left(-\frac{t}{\tau} \right) \right]. \tag{6.17}$$

Since Eq. (6.17) mimics Eq. (6.16) and since T_c measurements are sensitive to the OCA diffusion into the tissue, we can use those measurements to estimate the value of τ. As we have indicated in Sect. 6.2, biological tissues have different types of water, and during OC treatments, only the mobile water is able to move out from the tissue. We must consider that only for OC treatments where the water content in the treating solution matches the mobile water content in a tissue a unique OCA flux into the tissue will occur [30, 31]. Such unique flux is created due to the water balance between the tissue and the treating solution, and in that case only, τ in Eq. (6.17) matches τ_{OCA} in Eq. (6.16). For the particular case of a treatment with an oversaturated solution, due to the strong osmotic pressure that is created over the tissue, mostly water flows out from the tissue [29, 30]; and for that case, τ in Eq. (6.17) matches τ_{water}.

In general, the mobile water content is not known, and if we treat a tissue with a solution that has a different water content from that of the mobile water in the tissue, a mixed flux of water going out and OCA going into the tissue occurs. In this case, the τ value estimated with Eq. (6.17) corresponds to that mixed flux. A particular methodology is necessary to discriminate the τ values for water and for OCA in the tissue. By selecting a few OCA concentrations in the treating solution, a mean τ value can be estimated for each treatment, and representing these values as a function of OCA concentration in the solution, the dependence between both parameters can be obtained. Such representation is fitted with a smooth spline, where a maximum and a minimum will occur for an intermediate and an oversaturated solution, respectively [32, 33]. The maximum and minimum τ are the diffusion time values for the unique OCA and water fluxes, respectively [7, 30]. To show the sequence of these calculations, we will present as example the treatments of the rat skeletal muscle tissue with fructose solutions. This example was already discussed in Sect. 4.3, but now we will present the complete set of measurements and calculations.

We have treated skeletal muscle sample from the Wistar Han rat with water-fructose solutions that have the following fructose concentrations: 20%, 25%, 30%, 35%, 40%, 45%, 50%, 55%, and 60%. Those solutions were prepared according to the procedure indicated in Sect. 3.4. During each of these treatments, we measured collimated transmittance T_c spectra from the sample between 400 and 1000 nm. After terminating the experiments, we calculated the T_c time dependencies for wavelengths between 600 and 800 nm, the spectral region where the muscle tissue shows linear T_c increase and absence of absorption bands. Figure 6.12 shows these time dependencies.

Analyzing the various graphs presented in Fig. 6.12, we see a smooth increase at the beginning of treatment, with different extent for the different treatments. That smooth increase is the one that we need to fit with a curve described by Eq. (6.17), meaning that the following step is to trim the data at the end of the smooth increase. An additional step consists of displacing each curve to have $T_c = 0$ at $t = 0$ and normalizing the data in each displaced curve to its highest value. Figure 6.13 presents the data in graphs of Fig. 6.12 after displacement but without the normalization for better visual perception.

Graphs in Fig. 6.13 show the smooth behavior only. As we can see from these graphs, the extent of the smooth behavior increases until the treatment duration for treatments with low fructose concentration up to 40%. For treatments with higher concentrations, the extent of the smooth increase decreases with increasing concentration of fructose in the solution.

Considering each graph in Fig. 6.13, after normalizing each curve to the highest value, the data were fitted with a curve described by Eq. (6.17) to obtain the τ values. The mean τ values for each treatment and the corresponding standard deviation (SD) are presented in Table 6.1 [7].

Representing the mean τ values in Table 6.1 as a function of fructose concentration and interpolating those points with a smooth spline, one obtains the graph in Fig. 6.14.

By analyzing the graph in Fig. 6.14, one can retrieve valuable information. First, we see a maximum in the spline that occurs for a concentration of 40.6%. This means that the unique fructose flux into the skeletal muscle would occur for a solution with 40.6% of fructose. This concentration also provides the mobile water content in the muscle: 59.4%, a value that was already obtained in treatments with glucose and ethylene glycol (EG) [30, 31]. The τ value at this concentration is 314.2 s, and it corresponds to the diffusion time of fructose in the rat skeletal muscle.

The minimum τ value occurs for the highest concentration of fructose—60%. For this case, since the treating solution is oversaturated with fructose, the estimated τ value (57.7 s), which is very similar to others obtained with other OCAs [30, 31], corresponds to the diffusion time of water.

To obtain the correspondent diffusion coefficients, one needs thickness measurements to use in Eq. (6.13). We performed those measurements for the treatments with 40%- and 60%-fructose. Figure 6.14 presents these data.

The datasets in Fig. 6.14 were measured only once, but to eliminate experimental errors, these thickness measurements should be repeated two or three times [30, 31]. The curves in Fig. 6.14 are splines to interpolate the data points. Considering the diffusion time values obtained from Fig. 6.14, the corresponding sample thickness values were retrieved from the curves in Fig. 6.14 as $d_{musc\text{-}40\%\text{-}fructose}(t = 314.2\,s) = 0.419\,mm$ and $d_{musc\text{-}60\%\text{-}fructose}(t = 57.7\,s) = 0.404\,mm$.

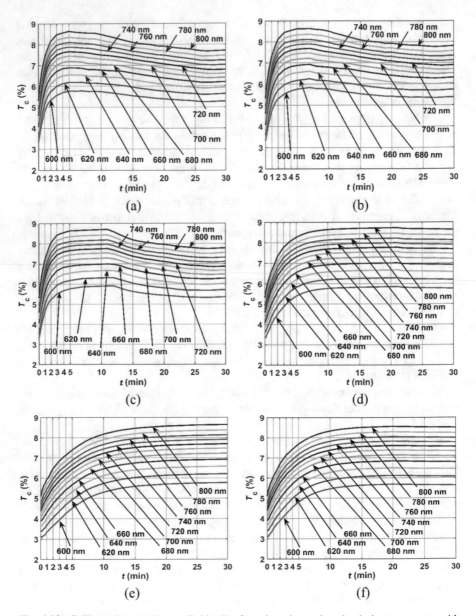

Fig. 6.12 Collimated transmittance T_c kinetics for selected wavelengths during treatments with solutions containing fructose in concentrations: 20% (**a**), 25% (**b**), 30% (**c**), 35% (**d**), 40% (**e**), 45% (**f**), 50% (**g**), 55% (**h**), and 60% (**i**)

Using these data in Eq. (6.13), the diffusion coefficients for fructose and water in the rat skeletal muscle tissue were calculated as $D_{\text{fructose-muscle}} = 5.7 \times 10^{-7}$ cm^2/s and $D_{\text{water-muscle}} = 2.9 \times 10^{-6}$ cm^2/s.

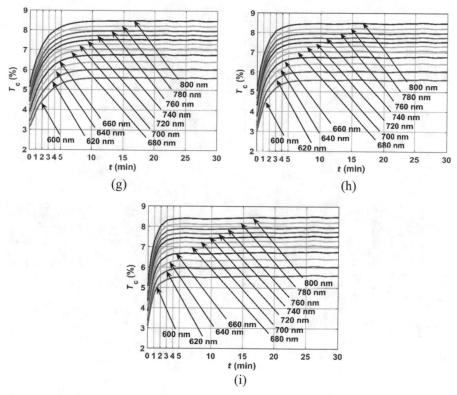

Fig. 6.12 (continued)

This robust method was used to estimate the diffusion properties of various agents in different tissues [7, 30–34]. Table 6.2 contains some of those values for D.

From Table 6.2, we see that water presents very similar diffusion coefficients in all normal tissues, while for pathological mucosa, it has a smaller value. On the other hand, we see that different OCAs have different coefficients in the same tissue—skeletal muscle. Considering glucose, we see that it shows different values in all cases presented in Table 6.2, but comparing between normal and pathological mucosa, we see that the coefficient is lower in the last case. Such difference is possibly related to the glucose different uptake in normal and pathological tissues, which is bigger for pathology. There are many other cases to be studied, and the method here described is a good tool to obtain those values, provided that studies are made for ex vivo samples (no any tissue fixation).

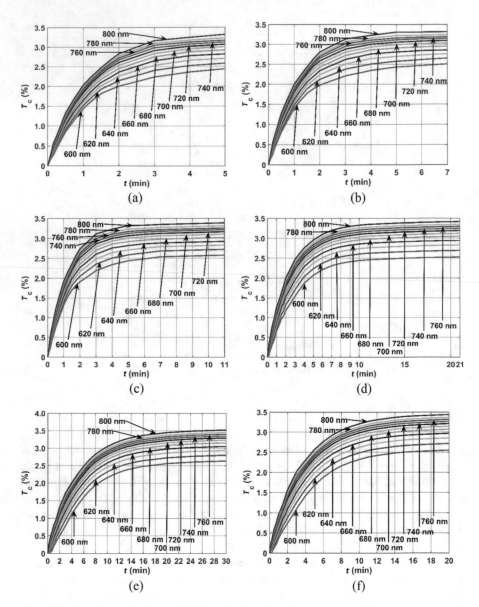

Fig. 6.13 Trimmed and displaced T_c kinetics for selected wavelengths during treatments with solutions containing fructose in concentrations: 20% (**a**), 25% (**b**), 30% (**c**), 35% (**d**), 40% (**e**), 45% (**f**), 50% (**g**), 55% (**h**), and 60% (**i**)

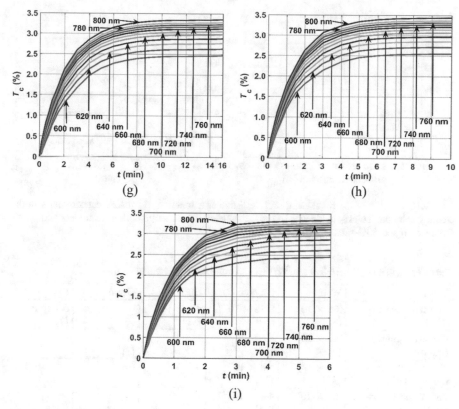

Fig. 6.13 (continued)

Table 6.1 Estimated τ values for the treatments of the rat skeletal muscle tissue with fructose solutions

Fructose (%)	20	25	30	35	40	45	50	55	60
Mean τ (s)	65.2	69.6	78.4	157.9	311.5	230.1	135.9	88.7	57.7
SD (s)	2.2	4.1	5.4	16.1	19.7	14.9	12.7	5.3	1.7

6.5 Physiological Data and Possibility of Diagnostic Procedures

In the previous section, we have described an ex vivo method to evaluate the diffusion properties of water and OCAs in biological tissues. This same method also provides information on the mobile water content in the tissues.

If a tissue has an unknown mobile water content, the above described method can be used to evaluate it. In fact, in the previous study [32], different mobile water content in human normal and pathological colorectal mucosa tissues was evaluated.

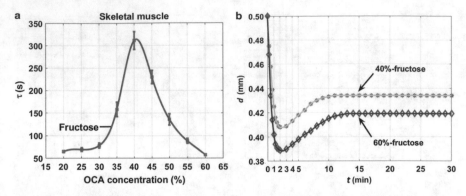

Fig. 6.14 Mean τ of the rat skeletal muscle tissue as a function of fructose concentration in the treating solution. Thickness kinetics of the rat skeletal muscle tissue during treatments with 40%-fructose (**a**) and 60%-fructose (**b**)

Table 6.2 Diffusion coeffcient (cm^2/s) of OCAs and water in biological tissues

Tissue	Water	Glucose	EG	Glycerol	PPG
Skeletal muscle (rat)	3.1×10^{-6} [31]	8.36×10^{-7} [30]	4.6×10^{-7} [31]	–	5.1×10^{-7} [7]
Colorectal muscle (human)	3.1×10^{-6} [7]	–	–	3.3×10^{-7} [7]	–
Normal colorectal mucosa (human)	3.3×10^{-6} [32]	5.8×10^{-7} [32]	–	–	–
Pathological colorectal mucosa (human)	2.4×10^{-6} [32]	4.4×10^{-7} [32]	–	–	–
Liver (human)	3.2×10^{-6} [33]	–	–	8.2×10^{-7} [33]	–

Similar studies were performed that allowed discrimination of the mobile water content from normal and diabetic rat skin and myocardium [36].

Figure 6.15 shows the τ values as a function of OCA concentration in the human colorectal tissues [7, 32].

The treatments of normal and pathological mucosa in Fig. 6.15 were performed with glucose-water solutions [32], while colorectal muscle samples were treated with glycerol-water solutions [7]. The graphs in Fig. 6.15 show that the τ has a peak at an OCA concentration of 35.6% for pathological mucosa, while for normal mucosa, the peak occurs at a concentration of 40.6%, and for muscle it occurs exactly at a concentration of 40%. Since the peaks are observed for an OCA concentration in a solution that contains the same amount of water as the mobile water in the tissue, we see that pathological mucosa presents about 5% more mobile water than normal colorectal tissues. Although this method is not adequate to use directly

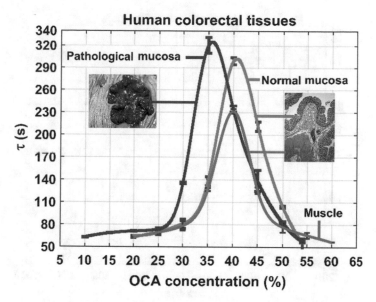

Fig. 6.15 Mean τ of the human colorectal tissues as a function of OCA concentration in the treating solution

in vivo, it provides valuable information and can be used as a complementary ex vivo diagnostic procedure. However, in vivo modification of this technique in the backreflectance mode is possible.

6.6   Creation of the UV-Tissue Windows

A new application of OC treatments to obtain physiological data and possible development of new diagnostic procedures results from recent research that was performed in the ultraviolet (UV) range. Until recently, the OC technique has been successfully applied to obtain results in the visible-NIR spectral range. Wondering what would be the effects of this technique in the UV range, a study on human colorectal muscle tissue under treatment with glycerol-water solutions with different osmolarities was done.

Treating tissues with 20%-, 40%-, and 60%-glycerol solutions, collimated transmittance T_c spectra were measured between 200 and 1000 nm. The T_c spectrum in this range for native colorectal muscle (sample thickness of 0.5 mm) is presented in Fig. 6.16.

From Fig. 6.16 we see that T_c increases almost linearly from 300 to 1000 nm, showing the Soret band at 415 nm. For shorter wavelengths, the T_c of muscle is almost null, suggesting that strong absorption and scattering occur in this range. It has been reported that strong absorption peaks occur at 200 nm for proteins and

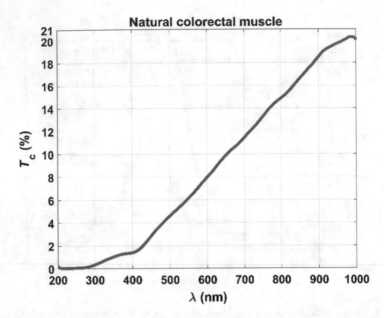

Fig. 6.16 Collimated transmittance T_c spectrum measured from native human colorectal muscle sample

at 260 nm for DNA [37]. The combination of these strong absorption peaks at short wavelengths turns muscle T_c very low between 200 and 300 nm.

Considering that the most significant mechanisms of OC are the RI matching and the dehydration mechanisms [4], the RI of tissue components, water, and OCAs increases with decreasing the wavelength [8, 38–42], and since the protein dissolution has also been reported [43], we should expect that the degree of increase in tissue transparency should be greater in the UV range.

By representing the kinetics of T_c spectra obtained during the treatments, we do not see the significant increase in tissue transparency in the UV range (Fig. 6.17) [44].

The graphs presented in Fig. 6.17 show typical kinetics for T_c of fibrous tissues: increasing magnitude of T_c increase in the visible NIR and very low increase in the UV range. However when T_c graphs are magnified in the UV spectral region, a very effective OC is well seen and can be quantified in the spectral range between 200 and 300 nm for all treatments.

From the graphs in Fig. 6.18, we see similar behavior in the UV range to what was observed in the visible-NIR range (Fig. 6.17): magnitude of the increase in T_c grows with glycerol concentration in the treating solution. In each of the magnified graphs of Fig. 6.18, we see also similar behavior on both sides of the DNA absorption band (central peak at 260 nm).

Considering that natural muscle shows low T_c between 200 and 300 nm, to make an accurate comparison between the UV and visible-NIR spectral ranges, instead of

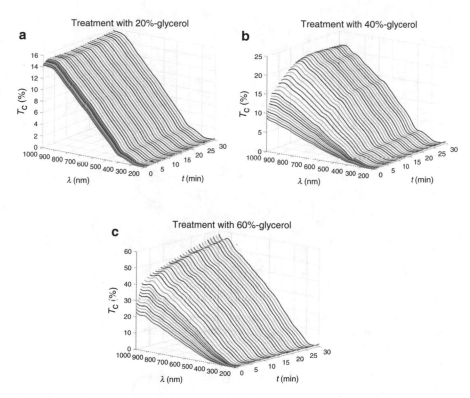

Fig. 6.17 Collimated transmittance T_c kinetics of the human colorectal muscle tissue during treatments with 20%-glycerol (**a**), 40%-glycerol (**b**), and 60%-glycerol (**c**)

comparing absolute variations in T_c, we need to analyze its rate (percent increase). Using the data in Fig. 6.17, we calculated the percent increase with Eq. (6.18):

$$\% \uparrow T_c = \frac{T_c(\lambda, t) - T_c(\lambda, t = 0)}{T_c(\lambda, t = 0)} \times 100\%. \tag{6.18}$$

The results of this calculation are presented in Fig. 6.19.

In all graphs of Fig. 6.19, we see a huge rise between 200 and 250 nm for all treatments. We see also an increase between 300 and 400 nm, which seems to be small for treatments with small glycerol concentration. The percent increase in both these ranges with glycerol concentration in solution is similar to what was observed in Fig. 6.17 for the visible-NIR spectral range, but the percent increase in UV is much more significant. Although all the graphs in Fig. 6.19 are represented for the wavelength range of 200–600 nm for better visualization, the percent increase at longer wavelengths is smaller.

Considering as an example the treatment of human colorectal muscle with 60%-glycerol (Fig. 6.19c), an increase of 260% is observed at 600 nm, while an increase of about 1300% is obtained at 230 nm.

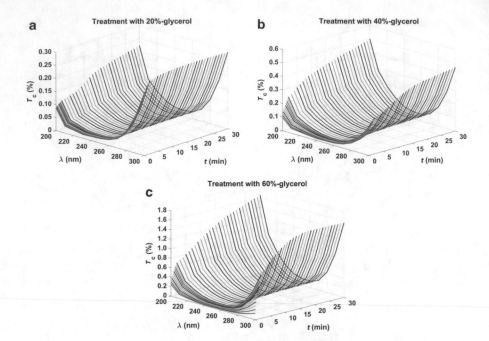

Fig. 6.18 Collimated transmittance T_c kinetics measured between 200 and 300 nm of the human colorectal muscle tissue during treatments with 20%-glycerol (**a**), 40%-glycerol (**b**), and 60%-glycerol (**c**) [44]

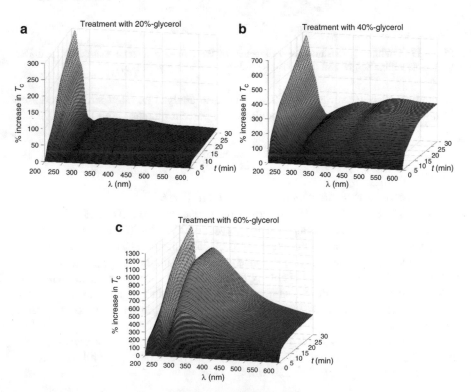

Fig. 6.19 Kinetics for the percent increase of T_c for the human colorectal muscle tissue during treatments with 20%-glycerol (**a**), 40%-glycerol (**b**), and 60%-glycerol (**c**) [44]

The graphs in Fig. 6.19 show the creation of two new optical windows showing the best rate of T_c increase in the UV range, which we designate hereafter as UV-1 and UV-2. The UV-1 window is created between 200 and 260 nm, with a central peak at 230 nm, while the UV-2 window is created between 260 and 400 nm, having its central peak at 300 nm.

These are two new and artificial optical windows that can be used for diagnostic and treatment purposes by using OC treatments. All these results are subject of a patent pending from authors of Ref. [44].

References

1. A. Kotyk, K. Janacek, *Membrane Transport: An Interdisciplinary Approach* (Plenum Press, New York, NY, 1997)
2. V.V. Tuchin, *Optical Clearing of Tissues and Blood* (SPIE Press, Bellingham, WA, 2006)
3. V. Hovhannisyan, P.-S. Hu, S.-J. Chen, C.-S. Kim, S.-Y. Dong, Elucidation of the mechanisms of optical clearing in collagen tissue with multiphoton imaging. J. Biomed. Opt. **18**(4), 046004 (2013)
4. L. Oliveira, M.I. Carvalho, E.M. Nogueira, V.V. Tuchin, Optical clearing mechanisms characterization in muscle. J. Innov. Opt. Health Sci. **9**(5), 1650035 (2016)
5. L. Silvestri, I. Constantini, L. Scconi, F.S. Pavone, Clearing of fixed tissue: a review from microscopist's perspective. J. Biomed. Opt. **21**(8), 081205 (2016)
6. A.Y. Sdobnov, M.E. Darvin, J. Schleusener, J. Lademann, V.V. Tuchin, Hydrogen bound water profiles in the skin influenced by optical clearing molecular agents – quantitative analysis using confocal Raman microscopy. J. Biophotonics **12**(5), e21800283 (2019)
7. I. Carneiro, S. Carvalho, R. Henrique, L.M. Oliveira, V.V. Tuchin, A robust ex vivo method to evaluate the diffusion properties of agents in biological tissues. J. Biophotonics **12**(4), e201800333 (2019)
8. L. Oliveira, M.I. Carvalho, E. Nogueira, V.V. Tuchin, Skeletal muscle dispersion (400-1000 nm) and kinetics at optical clearing. J. Biophotonics **11**(1), e201700094 (2018)
9. I. Carneiro, S. Carvalho, R. Henrique, L. Oliveira, V.V. Tuchin, Kinetics of optical properties of colorectal muscle during optical clearing. IEEE J. Sel. Top. Quant. Elect. **25**(1), 7200608 (2019)
10. C.-S. Choe, J. Lademann, M.E. Darvin, Depth profiles of hydrogen bound water molecule types and their relation to lipid and protein interaction in the human stratum corneum in vivo. Analyst **141**(22), 6329–6337 (2016)
11. I. Carneiro, S. Carvalho, V. Silva, R. Henrique, L. Oliveira, V.V. Tuchin, Kinetics of optical properties of human colorectal tissues during optical clearing: a comparative study between normal and pathological tissues. J. Biomed. Opt. **23**(12), 121620 (2018)
12. I. Carneiro, S. Carvalho, R. Henrique, L.M. Oliveira, V.V. Tuchin, Optical properties of colorectal muscle in visible/NIR range, in Biophotonics: Photonic Solutions for Better Health Care VI, ed. by J. Popp, V.V. Tuchin, F.S. Pavone. Proc. SPIE **10685**, 106853D (2018)
13. A.N. Bashkatov, E.A. Genina, V.I. Kochubey, V.V. Tuchin, Optical properties of human skin, subcutaneous and mucous tissues in the wavelength range from 400 to 2000 nm. J. Phys. D Appl. Phys. **38**(15), 2543–2555 (2005)
14. A.N. Bashkatov, E.A. Genina, V.I. Kochubey, V.S. Rubtsov, E.A. Kolesnikova, V.V. Tuchin, Optical properties of human colon tissues in the 350-2500 spectral range. Quant. Electron. **44**(8), 779–784 (2014)
15. S. Carvalho, N. Gueiral, E. Nogueira, R. Henrique, L. Oliveira, V.V. Tuchin, Comparative study of the optical properties of colon mucosa and colon precancerous polyps between 400 and

1000 nm, in Dynamics and Fluctuations in Biomedical Photonics XIV, ed. by V.V. Tuchin, K.V. Larin, M.J. Leahy, R.K. Wang. Proc. SPIE **10063**, 100631L (2017)

16. D.W. Leonard, K.M. Meek, Refractive indices of the collagen fibrils and extrafibrillar material of the corneal stroma. Biophys. J. **72**(3), 1382–1387 (1997)

17. K.M. Meek, S. Dennis, S. Khan, Changes in the refractive index of the stroma and its extrafibrillar matrix when the cornea swells. Biophys. J. **85**(4), 2205–2212 (2003)

18. K.M. Meek, D.W. Leonard, C.J. Connon, S. Dennis, S. Khan, Transparency, swelling and scarring in the corneal stroma. Eye **17**, 927–936 (2003)

19. O. Zernovaya, O. Sydoruk, V. Tuchin, A. Douplik, The refractive index of human hemoglobin in the visible range. Phys. Med. Biol. **56**, 4013–4021 (2011)

20. L.-H. Wang, S.L. Jacques, L.-Q. Zheng, MCML – Monte Carlo modeling of photon transport in multi-layered tissues. Comp. Met. Progr. Biomed. **47**(2), 131–146 (1995)

21. S.A. Prahl, M.J.C. Van Gemert, A.J. Welch, Determining the optical properties of turbid media by using the adding-doubling method. Appl. Optics **32**(4), 559–568 (1993)

22. R. Graaff, J.G. Aarnoudse, J.R. Zijp, P.M.A. Sloot, F.F. de Mul, J. Greve, M.H. Koelink, Reduced light-scattering properties for mixtures of spherical particles: a simple approximation derived from Mie calculations. Appl. Optics **31**(10), 1370–1376 (1992)

23. H. Liu, B. Beauvoit, M. Kimura, B. Chance, Dependence of tissue optical properties on solute-induced changes in refractive index and osmolarity. J. Biomed. Opt. **1**(2), 200–211 (1996)

24. E.A. Genina, A.N. Bashkatov, V.V. Tuchin, in *Handbook of Optical Sensing of Glucose in Biological Fluids and Tissues*, ed. by V. V. Tuchin, (CRC Press, Boca Raton, FL, 2009). Chapter 21

25. M.G. Ghosn, E.F. Carbajal, N.A. Befrui, V.V. Tuchin, K.V. Larin, Differential permeability rate and percent clearing of glucose in different regions in rabbit sclera. J. Biomed. Opt. **13**(2), 021110-1–021110-6 (2008)

26. Q.L. Zhao, J.L. Si, Z.Y. Guo, H.J. Wei, H.Q. Yang, G.Y. Wu, S.S. Xie, X.Y. Li, X. Guo, H.Q. Zhong, L.Q. Li, Quantifying glucose permeability and enhanced light penetration in ex vivo human normal and cancerous esophagus tissues with optical coherence tomography. Laser Phys. Lett. **8**(1), 71–77 (2011)

27. H. Ullah, E. Ahmed, M. Ikram, Monitoring of glucose levels in mouse blood with noninvasive optical methods. Laser Phys. **24**(2), 025601-1–025601-8 (2014)

28. O. Zhernovaya, V.V. Tuchin, M.J. Leahy, Blood optical clearing studied by optical coherence tomography. J. Biomed. Opt. **18**(2), 26014-1–26014-8 (2013)

29. P. Liu, Y. Huang, Z. Guo, J. Wang, Z. Zhuang, S. Liu, Discrimination of dimethyl sulfoxide diffusion coefficient in the process of optical clearing by confocal micro-Raman spectroscopy. J. Biomed. Opt. **18**(2), 20507-1–20507-3 (2013)

30. L.M. Oliveira, M.I. Carvalho, E.M. Nogueira, V.V. Tuchin, The characteristic time of glucose diffusion measured for muscle tissue at optical clearing. Laser Phys. **23**, 075606-1–075606-6 (2013)

31. L.M. Oliveira, M.I. Carvalho, E.M. Nogueira, V.V. Tuchin, Diffusion characteristics of ethylene glycol in skeletal muscle. J. Biomed. Opt. **20**(5), 051019-1–051019-10 (2015)

32. S. Carvalho, N. Gueiral, E. Nogueira, R. Henrique, L. Oliveira, V.V. Tuchin, Glucose diffusion in colorectal mucosa – a comparative study between normal and cancer tissues. J. Biomed. Opt. **22**(9), 091506-1–091506-12 (2017)

33. I. Carneiro, S. Carvalho, R. Henrique, L. Oliveira, V.V. Tuchin, Simple multimodal optical technique for evaluation of free/bound water and dispersion of human liver tissue. J. Biomed. Opt. **22**(12), 125002-1–125002-10 (2017)

34. I. Carneiro, S. Carvalho, R. Henrique, L. Oliveira, V.V. Tuchin, Water content and scatterers dispersion evaluation in colorectal tissues. J. Biomed. Phot. Eng. **3**(4), 040301-1–040301-10 (2017)

35. V.V. Tuchin, Optical immersion as a new tool for controlling the optical properties of tissues and blood. Laser Phys. **15**, 1109–1136 (2005)

36. D.K. Tuchina, A.N. Bashkatov, A.B. Bucharskaya, E.A. Genina, V.V. Tuchin, Study of glycerol diffusion in skin and myocardium ex vivo under the conditions of developing alloxan-induced diabetes. J. Biomed. Phot. Eng. 3(2), 020302-1–020302-9 (2017)

37. http://elte.prompt.hu/sites/default/files/tananyagok/IntroductionToPracticalBiochemistry/ch04s06.html. Accessed 2 Mar 2019

38. M. Daimon, A. Masumura, Measurement of the refractive index of distilled water from the near-infrared region to the ultraviolet region. Appl. Optics **46**, 3811–3820 (2007)

39. I.Z. Kozma, P. Krok, E. Riedle, Direct measurement of the group-velocity mismatch and derivation of the refractive-index dispersion for a variety of solvents in the ultraviolet. J. Opt. Soc. Am. B **22**, 1479–1485 (2005)

40. R.D. Birkhoff, L.R. Painter, J.M. Heller Jr., Optical and dielectric functions of liquid glycerol from gas photoionization measurements. J. Chem. Phys. **69**, 4185–4188 (1978)

41. E. Sani, A. Dell'Oro, Optical constants of ethylene glycol over an extremely wide spectral range. Opt. Mater. **37**, 36–41 (2014)

42. E. Sani, A. Dell'Oro, Corrigendum to optical constants of ethylene glycol over an extremely wide spectral range. Opt. Mater. **48**, 281 (2015)

43. J. Hirshburg, B. Choi, J.S. Nelson, A.T. Yeh, Collagen solubility correlates with skin optical clearing. J. Biomed. Opt. **11**, 040501 (2006)

44. I. Carneiro, S. Carvalho, R. Henrique, L. Oliveira, V.V. Tuchin, Moving tissue spectral window to the deep-UV via optical clearing. J. Biophotonics, e201900181 (2019)

Chapter 7
Optical Clearing and Tissue Imaging

7.1 Introduction

Since earlier times visual or microscopic inspection has played an important role in diagnostic medicine. With the development of new technologies, medical imaging has become an effective way to detect pathologies through visual analysis of the generated images. Traditional imaging methods, such as magnetic resonance imaging (MRI), computerized X-ray tomography (CT), or positron emission tomography (PET), are applied in many cases, but they are associated with some disadvantages.

Besides being expensive methods, they present some potential risks for patients with certain allergies or metal implants. Pregnant women cannot be submitted to the above imaging procedures to avoid damaging the fetus due to exposure to radiation or strong magnetic forces [1].

In face of these disadvantages, new low-cost and radiation-free imaging methods that can be used to acquire images from deep tissue layers are desired.

Several optical imaging techniques, such as optical coherence tomography (OCT), confocal microscopy, or fluorescence imaging modalities, are known for some time, but their application is restricted to low tissue depths, due to the high scattering properties in biological tissues. A solution to increase tissue depth is the combination of these imaging methods with the optical clearing (OC) technique [2]. With the reduction of light-scattering properties in biological materials, imaging methods can be applied more efficiently to acquire images from deeper tissue layers and with better contrast and resolution [2–4]. There are several imaging methods that have been used with the OC method to evaluate the improvement in the acquired images. The following sections address the low depth imaging problem from thick tissues and the various imaging methods where tissue clearing has provided improvement in tissue depth, contrast, and resolution.

© The Author(s), under exclusive license to Springer Nature Switzerland AG 2019
L. M. C. Oliveira, V. V. Tuchin, *The Optical Clearing Method*,
SpringerBriefs in Physics, https://doi.org/10.1007/978-3-030-33055-2_7

7.2 Tissue Imaging in Thick Tissue

Due to the strong scattering properties of biological materials, it is easier to image thin sections than thick samples. Traditionally, the preparation of two-dimensional tissue samples with the use of microtomes has been the conventional method to obtain images at the microscope with the purpose of establishing a diagnosis. Such technique can be used only on excised tissues and has no application for the in vivo examination [5].

Three-dimensional imaging is of great interest, since it provides more information, allowing for structural analysis of cells, tissues, and organs and identifying the occurrence of pathologies. Full organ in vivo imaging is also desired in some cases, since comparison between areas within the same organ can assist in establishing a diagnosis. Some techniques have been used that allow acquiring images from continuous deeper layers in a tissue, as each layer is excised after imaged. These are destructive techniques, but they allow 3D image reconstruction by combining the images from the various layers [5, 6].

Optical sectioning techniques, such as the laser scanning confocal microscopy or the light-sheet microscopy, have been developed as alternative and nondestructing methods to allow image acquisition from several thin sections within a tissue volume by minimizing contributions from other parts of that volume. The various section images can later be combined to create a 3D image [5].

Even using optical sectioning microscopy techniques, there are still major obstacles regarding imaging of thick tissues. The first one is the occurrence of some tissue components that absorb light. Many tissues contain hemoglobin, myoglobin, and melanin, which are the main absorbers in the UV and visible range [5]. Some tissues may also contain lipids and water, which present their absorption peaks in the near-infrared range (peaks at 762, 830, 930, and 1200 nm) [7]. The more important problem in thick tissue imaging is light scattering, which is created by the nature of the tissue and its internal composition. Light scattering turns tissues opaque, with milky appearance, which diminishes sharpness in images. Light scattering may be small for the top layers of a tissue, but as we progress down to deeper tissue layers, the effect is cumulative, and the image quality decreases progressively [5]. One example is fluorescence imaging from deep tissue fluorophores. Due to the cumulative scattering in the upper tissue layers, detected fluorescence shows small signal-to-noise ratio, leading to a degraded image quality. Induced fluorescence can also be a problem for some imaging modalities, such as Raman, since it can be created both by endogenous, such as the reduced form of nicotinamide dinucleotide phosphate (NADPH), collagen, flavins, and tyrosine or exogenous molecules, which are accidentally added during tissue fixation [5]. Some methods have been reported to reduce the background autofluorescence [8–10]. The OC method can be used to reduce light scattering and improve image quality, but special attention must be paid to the case of autofluorescence, since by reducing scattering, the photon path length and the cross section for photon interaction with fluorophore will also reduce [11–15], leading to a decreased autofluorescence signal.

With the exception of the induced autofluorescence problem, the OC method can be applied to improve tissue imaging. Several studies have been made to provide the improvement of image quality with the help of the OC method. The following sections present some examples for different imaging methods.

7.3   Optical Coherence Tomography (OCT)

OCT is an interferometric technique that uses coherent light and allows one to acquire tomographic data from slices within a volume of a specimen. This technique was first demonstrated in 1991, and the imaging version was successfully implemented with the acquisition of images from in vitro human retina and from atherosclerotic plaque [16–19].

The imaging version of OCT is similar to the one of ultrasound, since the intensity of reflected infrared light from the sample is measured in the first and reflected sound waves are acquired in the later. Due to the high light speed, to measure the echo time delay in the OCT technique, which is in the order of femtoseconds, an optical interferometer illuminated by a low-coherent light source is used [19].

The experimental setup to acquire OCT images uses a dual-beam Michelson interferometer, as represented in Fig. 7.1.

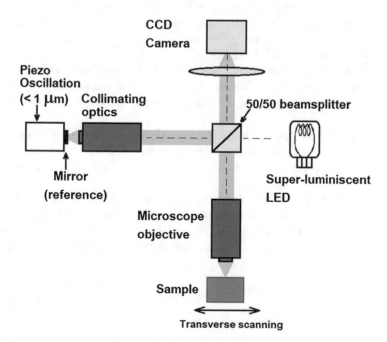

Fig. 7.1 Typical OCT imaging setup (adapted from Ref. [20])

The light beam from a super-luminescent LED (SLED) is separated into two beams in Fig. 7.1—the reference beam on the left and the sensing beam at the bottom. The reference beam is reflected by a mirror that can be moved through a piezoelectric transducer to allow interference with the sensing beam at a particular depth within the sample. The sample can be moved also to allow for transverse scanning. The reflected reference and sensing beams interfere at the 50/50 beam splitter, and the resulting interference signal is registered by the CCD camera on top.

When the mirror in the reference arm is moving with a constant speed v, the signal arising from the interference of the reflected reference and sensing beams is modulated at the Doppler frequency [19]

$$f_D = \frac{2v}{\lambda}. \tag{7.1}$$

The coherence length of the light source is defined as

$$l_c = \frac{2\ln(2)}{\pi} \frac{\lambda^2}{\Delta\lambda}, \tag{7.2}$$

where $\Delta\lambda$ represents the bandwidth of the light source with a Gaussian line profile. Due to the small coherence length of the light source, the Doppler signal is produced by backscattered light only within a very small region (in the order of the coherence length, l_c) that corresponds to the current optical path length in the reference arm. For a SLED with a bandwidth between 800 and 860 nm, the longitudinal resolution falls within the range between 5 and 20 μm. If, instead, a titanium sapphire laser with a wavelength of 820 nm is used, the bandwidth may reach 140 nm, and the corresponding resolution is 2.1 μm [19, 21].

As indicated in the previous section, OCT is also limited by the strong light scattering in biological tissues, which decreases contrast and spatial resolution and results in a small probing depth, around 2–3 mm for conventional OCT systems [19, 22].

To improve the probing depth and image quality, the tissue's light scattering can be temporarily reduced with the application of optical clearing agents (OCAs) [2, 3, 5, 22–24]. OCT images are a result of the reflected signal intensity measured at the depth z, from the tissue under study.

The signal intensity depends on tissue reflectance $\alpha(z)$ at tissue depth z and also on the total attenuation coefficient $\mu_t = \mu_a + \mu_s$ of the tissue. For surface layers of strongly scattering tissues, where the single scattering model is valid, the reflected power is proportional to $\exp(-\mu_t z)$, i.e., it can be approximated with the expression [25]

$$R(z) = A\exp(-\mu_t z) + B, \tag{7.3}$$

where A is a coefficient equal to $P_0\alpha(z)$, with P_0 being the optical power of the incident beam on the tissue surface, and $\alpha(z)$ is the tissue reflectance in the backward direction at depth z and B the background signal [24].

Since the scattering coefficient presents higher values than the absorption coefficient from the UV to the near-infrared range for many tissues [7, 19, 26–30], the scattering coefficient is the main responsible for the exponential decay of the ballistic photons in Eq. (7.3) [24]. As we have already referred in Chap. 3, the scattering coefficient depends on the refractive index (RI) mismatch between tissue components and fluids within the tissue volume. By raising the RI of the interstitial fluid, the RI mismatch decreases and the correspondent decrease in the scattering coefficient are recorded as the decrease of the slope of the OCT signal amplitude as a function of the probing depth [24, 31–35].

Some recent studies have demonstrated the improvement in OCT images by using the OC method. Most of these studies were performed in skin tissues, since the main objective of the OCT imaging is to acquire images in vivo, but other tissues have also been studied with this technique.

For example, a study was conducted to evaluate the improvement of OCT images from ex vivo bovine muscle samples under treatment with 40%-glucose solution [22]. During this study, the diffusion coefficient of glucose in muscle was also determined as $(2.98 \pm 0.94) \times 10^{-6}$ cm^2/s. As a result of the treatment, a fourfold increase in image contrast for objects located ~360 μm deep below the top surface of the sample was found. A significant increase of ~2.4-fold in the image contrast at a depth of more than 800 μm was also received.

Other studies were conducted on ex vivo stomach and on in vivo skin tissues after topical application of propylene glycol [36, 37]. The OCT images acquired in these studies indicate a 3.5-fold increase in the probing depth and show also improved contrast of the internal inhomogeneities. The topical application of glycerol or propylene glycol to the in vivo and ex vivo human skin has also been made to control light scattering in OCT imaging and allowed the observation of subepidermal cavity and malignant melanoma diagnostics [3, 38]. Moreover, differentiation between normal and pathological tissue can be obtained by the variation rate of the OCT A-scan slope in the course of OC [14, 39–41]. Regarding blood, using Doppler OCT in combination with OCAs allows for visualization of cerebral blood vessels and blood velocity determination in mice without damaging cranial bone [42].

An ex vivo study on human skin, where a combination of thiazone, polyethylene glycol 400 (PEG-400), and ultrasound was used to clear the tissues, showed an increase of 41.3% for the observation depth [43]. In another study, using a combination of 50%-glycerol and 50%-polyethylene glycol 400 (PEG-400) as an OCA was applied to clear skin in vivo [44]. In this study, ultrasound and dimethyl sulfoxide (DMSO) were used independently or combined as clearing enhancers, and it was demonstrated through OCT data that maximal clearing efficiency is obtained after 20-min treatment when both ultrasound and DMSO are used as clearing enhancers. For OCA application without ultrasound or DMSO as enhancers, no significant clearing effect has been observed within the treatment time.

The clearing potential of disaccharides has also been investigated in skin. An OCT imaging study was conducted to compare the OC efficiency of sucrose and maltose with the one of fructose on ex vivo rat skin [45]. This study showed that OC efficiency increases linearly with the OCA concentration in solution for all saccharides. Experimental results demonstrate that both sucrose and maltose have a better OC potential than fructose at the same concentration and sucrose is optimal, since it produces better increase in imaging depth and better signal contrast.

Another study has used a mixture of fructose, PEG-400, and thiazone (FPT) as an OCA to clear mouse dorsal skin [46]. By using OCT angiography in this study, it was demonstrated that the treatment with FPT leads to improved imaging performance for the deeper skin tissues, as a result of the induced skin dehydration and the RI matching mechanisms. The acquired OCT angiography images demonstrate enhanced cutaneous hemodynamics imaging performance with satisfactory spatio-temporal resolution and better contrast when combined with the FPT clearing treatment.

The comparison of different OC methods in skin has also been made using OCT imaging. A study has compared the reduction of skin scattering with three clearing procedures, where the evaluation of the optical detection depth (ODD) was made [47]. The first clearing procedure was performed by removing the epidermis by laser ablation, the second was made through topical immersion clearing with PEG-300 after preliminary laser ablation of the epidermis, and the third consisted only of topical immersion of intact skin with PEG-300. In tissues where skin surface was ablated, local edema has formed, leading to an immediate reduction of the ODD, but due to water evaporation from the damaged epidermis, the observed optical clearing effect over time is comparable to the one observed when PEG is applied to intact skin. This study also demonstrated the preliminary ablation of skin surface before application of OCA does not bring significant improvement in the ODD, when compared to the sole effect of ablation or topical OCA administration.

Another study has demonstrated the enhancement of the observation depth between 300 and 750 μm for melanoma microvasculature through OCT with the application of PEG-400 and 1,2-propanediol solution (in proportions of 19:1) [38]. Also using OCT, it was showed that intradermal injection of fructose saline solution combined with intravenous injection of PEG-300 creates a fast clearing effect in mice in vivo, while injection of OCAs only into blood vessels does not provide any significant clearing effect [14, 48].

Considering cartilage tissues, it was demonstrated that the application of Omnipaque™ allows for observing OCT depths up to 0.9 mm, when the clearing effect saturated after 50 min [49]. Another study performed on cartilage during treatments with glucose solutions showed that by increasing glucose concentration in the treating solution causes changes in cartilage stiffness [50]. In this study, by applying glucose solutions with concentrations of 30%, 40%, and 70%, cartilage stiffness decreased by 44%, 55%, and 76%, respectively. The authors of this study suggested the water replacement by glucose in the extracellular matrix and partial cartilage dehydration is responsible for the observed cartilage stiffness decrease.

Other physical properties of tissue can also change with OC treatments, as demonstrated with the OCT imaging method for fresh ex vivo chicken breast, where topical application of glycerol solution has decreased tissue temperature and reduced its electro-kinetic response [51].

7.4 Microscopy Methods

The microscope has been used for many years to look into small details of several materials. Its application in clinical practice for cell visualization is also known since Middle Ages. Current microscopy technology has many variants that can be used on ex vivo or on in vivo tissues. Some of those variants are confocal microscopy, phase contrast microscopy, or fluorescence microscopy.

A typical fluorescence microscopy scheme is presented in Fig. 7.2.

In Fig. 7.2, a broadband light source is used to illuminate the sample. A filter selects the excitation wavelength, which is reflected in a 90° angle to hit the sample. This beam stimulates the fluorescence emission at the sample, and the fluorescence beam can be viewed at the ocular or detected by a CCD camera placed on top to acquire the images.

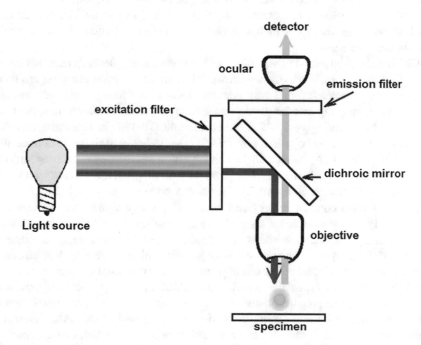

Fig. 7.2 Typical fluorescence microscope setup (adapted from Ref. [52])

Due to the discovery of the immersion OC method, many research studies have been recently performed using the various microscopy modalities. These techniques can be applied ex vivo or in vivo to image the skin.

Confocal microscopy (CM) is an optical imaging technique widely used to visualize the internal structure of biological tissues on a cellular and subcellular level [14, 53, 54]. In this technique, the detected light can be of scattering or of fluorescence nature, and it originates within a small volume of the sample, allowing to obtain high-quality images with high contrast and a micron level spatial resolution. The main limitation of the CM technique is the resulting low contrast images due to light scattering, but the application of the immersion OC technique can improve such low contrast, spatial resolution, and also the probing depth of the incident light [14].

Owing to OC of thick tissue sections, a technique was introduced in 2006 to resolve the microvasculature of murine tissues through CM of up to 1500 μm below the specimen surface [55]. By using FocusClear as an OCA, 3D mapping of entire brains of cockroach *Diploptera punctata*, which measures more than 500 μm in thickness, was possible [56]. FocusClear has also been used as an OCA on 1 mm—thick samples of mouse brain to perform optical histology and evaluate the OC potential [57]. In this study, the samples were also submitted to measurements before and after the treatment to estimate the difference in the optical properties of mouse brain as a result of the treatment. Similar variations were observed between 400 and 1000 nm, and the mean optical clearing potential has increased with the time of treatment. Confocal micrographs of mouse colon and ileum show that FocusClear can be used to reduce the opacity of these tissues and increase image resolution [58]. It should be noticed that studies reported in Refs. [55–58], tissue fixation with 4%-paraformaldehyde was made prior to measurements.

CM is also widely used to visualize gene expression and developmental patterns in plants. Once again, due to the significant RI mismatch in thick plant tissues, their imaging can be challenging. Many common OCAs are incompatible with protein fluorescence, meaning that for plant visualization, the OCA selection must be done carefully. Due to the success of 2,2′-thiodiethanol (TDE) in clearing animal tissues for fluorescence microscopy imaging [59–61], Musielak et al. have performed the evaluation of this OCA, through CM and two-photon microscopy in clearing various tissues from the *Arabidopsis thaliana* plant [62]. In this study, solutions with OCA concentrations between 50% and 70% showed a good compromise between tissue clearing for good contrast images and good fluorescence signal from proteins. A similar study on tissues from the same plant after clearing also with TDE has allowed for visualizing fluorescence stains at a depth of 200 μm and analyzes internal proteins of plant organs with confocal laser scanning microscope [61]. Visualization from the adaxial to abaxial epidermis of plant leaves as well as of the protoxylem and metaxylem vessels of vascular bundles embedded in spongy cells was possible. Without the need to prepare flower sections or remove sepals, inner floral organs were also observed from flower buds after clearing with TDE. After clearing, multicolor images of fluorescent proteins and dyes were obtained, and analyses of the three-dimensional structure of plant organs based on optical sections were possible. Root knots were also analyzed after clearing and revealed that nematodes induce giant cell expansion in a DNA content-dependent manner [61].

Visualization of the fine neuron structure in the inner regions of thick brain specimens allows for clarification of neural circuit functions. Also in this case, light scattering is a problem for the tissue penetration depth of laser scanning microscope. Aoyagi et al. have proposed a clearing protocol for mouse brain slices, using TDE solutions, for rapid enhancement of the penetration depth, both in confocal and in two-photon microscopy [59]. This research has demonstrated that the 30-min treatment of 400 μm thick brain slices allows to visualize dendritic spines along single dendrites at deep positions in the samples. A similar study was performed by Pavone's group, where the effectiveness of TDE in fixed whole mouse hippocampus brain samples was evaluated [60]. This study demonstrated that every individual part of the hippocampus could be imaged in high resolution, allowing one to resolve spines and varicosities across the whole area. By combining the acquired 2D images, it was possible to trace single neuronal processes with high accuracy throughout the hippocampus. According to the authors of Ref. [60], TDE is a versatile clearing agent for imaging methods in the brain and other anatomical areas, since it presents less viscosity than others such as SeeDB, is water soluble, preserves tissue fluorescence, and is cheaper than FocusClear.

CM images taken from various murine tissues revealed excellent transparency and preservation of cell morphology and protein and organic fluorescence after treated with the designated clearing-enhanced 3D (Ce3D) [63]. The Ce3D is a clearing medium that is prepared by diluting melted N-methylacetamide to 40% (v/v) in phosphate buffered saline (PBS), which is then used to dissolve Histodenz to 86% (v/v) concentration inside a chemical fume hood, with the mixture incubated at 37 °C to accelerate dissolution time. At the end, Triton X-100 (0.1% v/v) and 1-thioglycerol (0.5% v/v) are added to the solution. The final clearing medium presents a RI between 1.49 and 1.50. Stained and washed tissues from various organs (murine species) were placed in the Ce3D medium inside a chemical fume hood and incubated at room temperature on a rotor for 12–72 h, depending on tissue size [64]. The acquired images permitted quantitative analysis of distinct and highly intermixed cell populations via 3D histo-cytometry of high-volume specimens. Although good results were obtained with this protocol, they were obtained from ex vivo tissues that were previously fixed. For future in vivo imaging protocols, one has to avoid tissue fixation and be careful about OCA and enhancer selection to avoid induced tissue toxicity.

The investigation of saline, pure glycerol, and 80% DMSO solutions on dorsal mouse skin has also been performed with confocal microscopy [65]. By measuring images at 488 nm, it was observed a significant increase in skin scattering anisotropy and little change in the scattering coefficient, while DMSO produced insignificant variations [14]. It has been suggested that the induced glycerol clearing effect starts when reducing the angular deviation of scattering and an increase in anisotropy with minor change in the scattering coefficient should cause an increase in the scattering particle size, meaning swelling of collagen fibers in dermis [14].

Quantification of the clearing impact on cells, collagen, proteoglycans, and minerals in cartilage and bone tissues from the articulating knee joint has also been investigated with confocal microscopy imaging [66]. In this study, water-fructose solutions with different osmolarities were used to clear bovine osteochondral tissues, resulting in enhanced light transmission through cartilage, but not in subchondral bone regions. For cartilage tissues, light transmission was confirmed for 2.5 mm-thick tissues after clearing, and since sample preparation did not alter the fluorescence labeling, it was possible to acquire images from cartilage depths five times higher than for untreated tissues. Such images showed that tissue morphology remained unchanged and structural integrity was preserved during the clearing treatment. Complementary mechanical evaluation showed that cartilage treated tissues increased their equilibrium modulus, which returned to control levels after clearing reversal. These variations were attributed to the exchange of interstitial fluid's water by the more viscous fructose. For cartilage tissues that were not cleared (control), a significant decrease in the equilibrium modulus was observed [66].

As we can understand from the recent research presented in previous paragraphs, the application of tissue OC in imaging methods should provide multi-scale high-resolution images with tissue integrity preservation for accurate signal reconstruction. To avoid compromising tissue structure and maintain fluorescence signal stability, the selection of OCAs and protocols for imaging purposes should be done carefully. Considering this fact, some groups have developed their own clearing agents and protocols, and three of these cases are reported next.

Imaging of neural circuits from intact flies is limited by the structural properties of the cuticle. To overcome this problem, Dodt's group has developed an optical clearing protocol, designated as FlyClear, which is a signal preserving technique that stabilizes tissue integrity, efficiently removes overall pigmentation, and maintains fluorescence signal intensity for over a month [67]. The basic clearing solution used in this protocol is prepared by mixing 8 wt% THEED (2,2′,2″,2‴-(Ethylenedinitrilo)-tetraethanol), 5 wt% Triton® X-100, and 25 wt% urea. Applying different variations of the clearing procedure to the fruit fly, *Drosophila melanogaster*, from the larval to the adult stage, images obtained with ultramicroscopy, light-sheet and laser scanning confocal microscopy techniques allowed them the study of neural networks with single-cell resolution. From the light-sheet images, long-range neural connections were visualized from the peripheral sensory and central neurons in the visual and olfactory system [67].

ScaleS, a sorbitol-based clearing method that provides three-dimensional (3D) signal rendering and stable tissue preservation for immunochemical labeling, was recently reported [68]. The ScaleS solution is composed of 9.1 M urea and 22.5%-sorbitol, and depending on the desired transparency increase, some concentrations of up to 5% of Triton X-100 were added, producing solutions with designations according to Triton concentration (ScaleSQ(0) corresponds to 0%-Triton concentration and ScaleSQ(5) corresponds to a concentration of 5%). These solutions were used to treat thick mouse and human brain tissue samples, with the objective of comparing between different ages and between normal and Alzheimer's disease. The human samples were retrieved postmortem from patients that suffered from Alzheimer disease, with ages between 60 and 80 years old. The applied treatments permitted optical reconstructions of aged and diseased samples, including mapping of 3D networks of amyloid plaques, neurons, and microglia and multi-scale tracking of single plaques by successive fluorescence and electron microscopy. Comparison with results obtained with other clearing solutions showed superior signal and structure preservation by ScaleS. Using transmission electron microscopy to examine synaptic ultrastructure in treated samples, thin sections were excised from a 1 mm cube of treated sample after restoring to its original state by washing in PBS. Membrane integrity and ultrastructure preservation was observed, as well as subcellular organelles in the neurons. Identification of a symmetric synapse type (presumed inhibitory) in the axon initial segment and asymmetric synapses (presumed excitatory) in the neuropil region was possible. ScaleS treatments enabled high resolution of antibody epitopes for quantitative 3D immunohistochemical imaging of amyloid-β plaques in an Alzheimer's disease mouse and in Alzheimer's postmortem patient brain [68]. Another recent study on brain tissues from mice with Alzheimer's disease has reported a significant increase of amyloid-β plaques and the decrease of neurons in diseased tissues [69]. In this study, two optical clearing mixtures were used. The first one (OCA1) contained 25 wt% of urea, 25 wt% of meglutamine, 20 wt% of 1,3-dimethyl-2-imidazolidinone, and 0.5 wt% Triton X-100. The second (OCA2) contained 25 wt% of urea, 20 wt% of 1,3-dimethyl-2-imidazolidinone, and 40 wt% of sucrose. Each tissue slice was initially immersed in 5 mL of a water dilution of OCA1 (50%) at room temperature for 30 min and then immersed in 5–10 mL of OCA1 (100%) for 2 h. The treated samples were washed out with PBS several times and then incubated with a primary antibody—mouse polyantibody to amyloid beta (1:100) for 2–3 days at 37 °C. After this incubation, the samples were washed with 0.2% Triton/ PBS for 6 h and then incubated with the secondary antibody—goat polyantibody to mouse IgG conjugated to Alexa 555 (1:200) for 2–3 days at 37 °C. Following the second incubation, the samples were once again washed with 0.2% Triton/PBS and then treated with OCA2, before submitted to confocal imaging. Using an inverted laser scanning confocal microscope system with 10× and 20× objective lens, 3D fluorescence images were acquired from the samples through a self-built optical projection tomography system. A substantial increase in amyloid-β plaques with corresponding decrease in the number of neurons surrounding the plaques was observed in the acquired images for the diseased sample areas. For the healthy areas within the samples, where no amyloid-β plaques are found, the neurons were regular and free of deformation [69].

7.5 Speckle Techniques

The laser speckle contrast imaging (LSCI) was first introduced in the early 1980s [14, 70], and it is a noninvasive, low-cost, and easy to use method with great application to acquire subcutaneous and blood flow images [71–76]. Figure 7.3 presents a typical schematic for the LSCI setup.

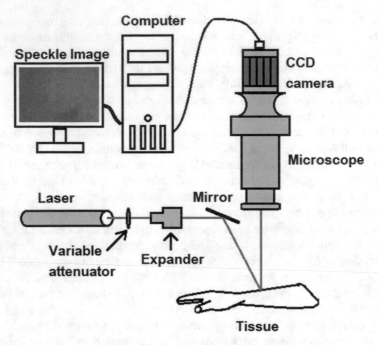

Fig. 7.3 Laser speckle imaging setup (adapted from Ref. [77])

Considering Fig. 7.3, a visible laser (typically emitting at 532 nm or at 632.8 nm) is used to illuminate the sample tissue. A variable attenuator is used to control beam power before being expanded to a diameter ~1 cm. To avoid intensity enhancement by coherent light backscattering, the incident beam is reflected by a mirror at a small angle with respect to the detection path. The backscattered beam is collected by a microscope, and images can be acquired via CCD camera [77].

If any parts of the specimen under study with the LSCI method are moving, temporal fluctuations in the single speckle intensity are translated into blurring of the speckle pattern during observation with a fixed camera exposure duration. Such blurring leads to a contrast reduction, but a statistical analysis of the blurred degree can provide information about the movement in the specimen. This technique has been used to acquire high-spatial and high-temporal resolution images of the 2D blood flow distribution in humans and animals [78]. The possibility of using LSCI to monitor the influence of OCAs in the blood flow is of high interest for researchers these days.

A decrease in blood flow as a result of topical application of glucose on mesenteric microvessels was first demonstrated by authors of Ref. [79]. It was observed that blood flow decreased from 1075 to 202 μm/s in venule with 11 μm diameter after 5 s of OCA interaction. Blood stasis was observed in the venule after 20 s of glucose interaction and after 20–30 s of glycerol interaction [79].

Another study has showed that direct topical application of glycerol and glucose on blood vessels located in the chick chorioallantoic membrane can decrease the blood flow velocity and also block the vessels [80]. In this study, the short-term observation showed that glucose induces small vessel blockage and that larger vessels only block after long-term interaction. Glycerol, on the other hand, caused similar damaging effects but much faster than glucose.

The time between the beginning of the OC treatment and the observation of skin vessels and blood flow should be kept to a minimum, since dermal vessels become invisible very quickly due to the fast contrast decrease after treatment. Zhu's group has developed an OC method that uses a mixture of PEG400 and thiazone to maximize the clearing effect in a short time, which allows imaging the dermal blood flow through the skin after 40 min of OCA treatment [81, 82]. Using this method in rat skin, it was demonstrated that after OCA application, speckle contrast improves and more data related to the dermal blood vessels or blood flow can be obtained.

Reversibility of the OC effects on blood is also of interest. Zhu's group has investigated the influence of 30%-glycerol solution and its reversibility on the dermal blood vessels of flap window of rat skin using LSCI [83]. This study demonstrated that blood flow decreased initially and began to recover after 16 min, indicating the reversibility of the clearing effect.

Skin microvascular visualization through the LSCI method can also provide information about pathologies. A recent paper has reported the microvascular dysfunction of type 1 diabetes in mice [84]. In this study, monitoring of the noradrenaline-induced responses of vascular blood flow and blood oxygen upon the development of diabetes could be made with a combination of LSCI, hyperspectral imaging, and optical clearing treatments. A decrease without recovery was found in this study for venous and arterious blood flow after injection of noradrenaline, and the oxygen decrease in arterious blood induced by noradrenaline weakened, especially for mice that had type 1 diabetes for more than 2 weeks. This study demonstrated that visualization of skin microvascular function is a potential research biomarker for early warning in the occurrence and development of diabetes and possibly other pathologies [84].

An innovative transparent cranial window for LSCI has been created on mouse skull by application of OC treatment and allowed to enhance the contrast of both white-light and speckle images from the microvessels in the cortex [85]. Using in vitro skull samples at first, complete transparency of the skull has been observed 25 min after OCA application, and images of the target placed underneath the skull showed a resolution of 8.4 μm. For the in vivo skull, topical application of the OCA was performed, leading to clear visualization of the cortical microvessels. Analysis of the LSCI images showed that the blood in the cortical microvessels was not affected by the OCA applied to the skull and a minimum resolution diameter for cortical microvessels of 12.8 ± 0.9 μm was obtained.

The study of blood microcirculation in the pancreas of rats with diabetes was also studied with LSCI after application of Omnipaque™ [86]. This study showed that the disease development causes changes in the response of the microcirculatory system due to OCA application. An increase of blood flow was observed in pancreatic vessels only in diabetic mice, but without change of the vessel diameter.

The study of cerebral blood flow was also studied in newborn mice using LSCI [76, 87]. Using aqueous 60%-glycerol, aqueous 70%-Omnipaque™ (300), and Omnipaque™ (300) solution in water/DMSO (25%/5%) to treat mice head without scalp removal and skull thinning, LSCI was used to acquire images from tissues in the area of the fontanelle. The analysis of the acquired images shows that the clearing effect is more efficient for the treatment with the aqueous solution containing 60% of glycerol but a reduction of the blood flow was also observed for this treatment: 12% decrease without skin removal and 40% after skin removal. Comparison between the results obtained from the treatments with the three solutions shows that solutions containing Omnipaque™ affect less the blood circulation both before and after skin removal. The optimal clearing efficiency was obtained for the treatment with the solution containing DMSO and Omnipaque™ [87].

Tuchin's group has recently conducted a study on pancreatic microcirculation under OC [88]. The results obtained with laser speckle contrast imaging (LSCI) were compared between normal and diabetic rat pancreas under treatment with solutions containing glycerol and PEG-300. It has been observed that application of 70%-glycerol demonstrates 50% decrease of blood flow velocity for diabetic animals, while almost 100% decrease of blood flow was observed for the control animals. For animals treated with PEG-300, a 25% decrease of blood flow velocity was observed for animals in the diabetic group, and a 65% decrease was obtained for the control group [88]. Such differences indicate that the combination of LSCI with OC is useful for pathology differentiation by evaluation of blood flow. Similar studies can be applied in other organs and to evaluate other pathologies.

A study has been reported by Abookasis and Moshe where a hexagonal lens array was added to a standard LSCI system to allow obtaining multiple speckle contrast map projections [89]. Using first a plastic tube filled with scattering liquid running at different controlled flow rates hidden in a biological tissue, the instrumental setup was validated. Once validated, this instrumental setup was used to image cerebral blood flow in intact rodent heads under hypoxic ischemic brain injury (group 1, $n = 5$), anoxic brain injury (group 2, $n = 5$), and functional activation (group 3, $n = 5$). OC treatments with a solution composed by liquid paraffin and glycerol were applied to increase tissue transparency and penetration depth. For all studied groups, the experimental results show the reduction in blood flow has a result of glycerol interaction. They also show improvement in the detection of blood flow and in signal-to-baseline ratio as a result of the introduction of the lens array [89]. This same group has also showed that an increase in temperature from 22 to 28 °C leads to an increase in contrast-to-noise ratio [90]. A more recent improvement of this technique involves the use of polarized light, which allows further improvement of image contrast. The creation of elliptical polarized speckle contrast projections with use of glycerol/liquid paraffin to clear chicken breast tissue has allowed to clearly visualize at different depths of a tumor mass (prostate cancer cells), which was introduced artificially [91].

The evaluation of the optical clearing effects of three sugars (fructose, glucose, and ribose) for imaging skin vessels at high contrast and resolution through dynamics simulation in vitro and in vivo studies has been reported [92]. At the beginning of the experiments and to evaluate the clearing potential of the sugars, fresh dorsal skin samples from Sprague Dawley rats were prepared to be used in the in vitro experiments. A total of 27 samples were prepared with a square form (1 cm side) and thickness of 620 ± 32 μm and then divided in three groups to be treated with fructose, glucose, and ribose solutions. After 24-min treatment, the samples were placed over a 1951 USAF target, and the treatments with fructose presented the optimal clearing potential. For the in vivo studies, the LSCI technique was used to monitor the cutaneous vascular structure and function, including imaging resolution, contrast, and sensitivity of the blood flow's dynamical response to vasoactive drug [93]. After treatment with fructose, PEG(400)-thiazone mixture, or fructose-PEG (400)-thiazone mixture for 30 min, the minimum resolution diameter of cutaneous blood vessels achieved 111.8 ± 6.1 μm, 49.6 ± 4.8 μm, or 35.8 ± 2.9 μm, respectively, meaning after treatment with fructose-PEG(400)-thiazone mixture, structure and functioning of smaller microvessels can be distinguished. Additionally, for the dermal blood vessels in the size range of 80 μm, the contrast-to-noise ratio of the speckle contrast images has improved 204% for the treatment with fructose-PEG (400)-thiazone mixture and 32% for the treatments with fructose and PEG(400)-thiazone [92].

7.6 The Second Harmonic Generation (SHG) Method

The application of second harmonic generation (SHG) imaging to biological materials was first performed in the 1980s, to study the orientation of collagen fibers in rat tail tendon [94]. Although microscope images with low resolution of ~50 μm were acquired in this study, it was the precursor of modern SHG imaging.

With the creation of new technology advances, such as ultrafast lasers, mode-locking, laser scanning, and data acquisition hardware/software, the SHG technique has evolved into a feasible tool for biological imaging. Various research studies were reported between 2000 and 2010, regarding visualization of collagen fibers in several connective tissues, in fibrous tissues from various internal organs, in cornea and skin tissues, or even in blood vessels. With these studies, the SHG technique has emerged as a powerful nonlinear contrast method for biological and biophysical applications. In the SHG imaging method, two lower energy photons are absorbed by the specimen to provide the subsequent emission of one photon with exactly twice the incident frequency (or half the wavelength) [95]. The occurrence of such nonlinear second-order optical process requires a nonsymmetric media, e.g., an anisotropic crystal, or an interface such as a membrane [19]. By stimulating the emission of double-frequency photons, the SHG technique presents some advantages for tissue imaging, such as allowing to easily reject surface reflection and multiple scattering of the incident light in biological tissue layers, such as in the epidermis layer in skin [19].

The total polarization of a material can in general be described by [94, 95]:

$$P = \chi^{(1)}\vec{E} + \chi^{(2)}\vec{E}\vec{E} + \chi^{(3)}\vec{E}\vec{E}\vec{E} + \cdots, \qquad (7.4)$$

where P is the induced polarization, $\chi^{(n)}$ is the nth order nonlinear susceptibility, and $\vec{E}$ is the electric field vector of the incident light. Severe symmetry restrictions on the harmonophores and their assembly that can be imaged are imposed by the second-order nature of SHG, since the environment must not have symmetry at the size scale of λ_{SHG} [94]. The first term in Eq. (7.4) concerns normal absorption and reflection of light, while the second term regards SHG and sum and difference frequency generation and the third term regards two- and three-photon absorption and third harmonic generation [19].

Since the SHG process is coherent, the SHG pulse is temporally synchronous with the excitation pulse, and the bandwidth scales are defined as $1/\sqrt{2}$ of the bandwidth of the excitation laser. The SHG signal intensity can be defined as [94, 95]:

$$I(2\omega) \propto \left[\chi^{(2)}\frac{E(\omega)}{\tau_p}\right]^2 \tau_p, \qquad (7.5)$$

where $E(\omega)$ and τ_p represent the laser pulse energy and width, respectively. Equation (7.5) shows a quadratic signal with peak power, but due to the instantaneous SHG process, the signal is generated only during the time extent of the laser pulse [94]. By measuring $I(2\omega)$ and knowing both $E(\omega)$ and τ_p, the calculation of the second-order susceptibility $\chi^{(2)}$ is performed. Although $\chi^{(2)}$ represents a bulk property, it can be expressed in terms of the first hyperpolarizability, β, which is a molecular property of nonlinearity [94, 95],

$$\chi^{(2)} = N_s\langle\beta\rangle, \qquad (7.6)$$

where the molecular density is represented by N_s and the brackets denote their orientational average, which emphasizes the need for an environment lacking a center of symmetry. The contrasting mechanism depends on β, which can be defined in terms of the permanent dipole moment, d [94, 95]:

$$d^{(2)} = \beta EE. \qquad (7.7)$$

Due to this proportionality between β and d, the alignment of the various dipoles provides maximization of the measured $\chi^{(2)}$ and consequently of the measured signal intensity, I (Eq. 7.5). This means that maximal SHG contrast will be obtained for well-aligned molecules that assembly into fibrils. When analyzing supramolecular structures, deviations from ideal alignment can be detected through decreased $\chi^{(2)}$ values [94].

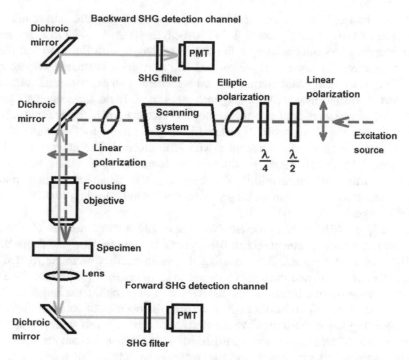

Fig. 7.4   SHG imaging microscope setup (adapted from Ref. [96])

A typical setup for the SHG microscope is presented in Fig. 7.4, where the excitation beam passes through a half-wavelength ($\lambda/2$) and a quarter-wavelength ($\lambda/4$) plates before being polarized elliptically.

In this setup that contains both forward and backward detection channels, the excitation source is normally a Ti:sapphire laser, which emits pulses in the range of 80–200 fs width at a repetition rate between 50 and 100 MHz. The emission power for this laser can reach 5 W for a tunable beam in the range between 700 and 1000 nm, which corresponds to the range where biological molecules have a significant cross section for SHG [97]. Such laser characteristics make the Ti: sapphire laser suitable for imaging a few hundred microns of tissue thickness. The excitation beam passes through the scanning system, which is connected to the photomultiplier tubes (PMT) in the forward and backward channels to maximize signal intensity. In the configuration presented in Fig. 7.4, both transmitted and reflected SHG images can be obtained from the specimen.

The SHG imaging technique has been used with optical clearing, and some studies have been reported.

A study has been reported by Campagnola's group for muscle clearing with 25%-, 50%-, and 75%-glycerol solutions [97]. Each muscle cell in striated muscle is wrapped around by a collagenous endomysium layer (1–2 μm thick). The RI mismatch at these interfaces between muscle cells ($n = 1.38$) and collagen ($n = 1.47$) originates strong light scattering. Making some calculations, when applying the treatments with the above glycerol solutions, the intracellular RI increases to 1.40 (25%-glycerol), 1.43 (50%-glycerol), and 1.45 (75%-glycerol). Then these authors calculated μ_s' at the SHG wavelength with Eq. (6.9) and found that it decreases by factors of ~2, 6, and 25, after each treatment, respectively [97]. Such calculated variations are in good agreement with experimental variations obtained for murine muscle [98]. Using such experimental data, Campagnola's group performed some simulations to estimate the SHG conversion efficiency for the above glycerol treatments, but a twofold greater efficiency was found for the treatment with 25%-glycerol, relative to the treatment with 75%-glycerol. The authors explained these results by the 50–100% swelling of the tissue during treatment that causes increased spacing of the myofibrils. Consequently, after treatment, harmonophores from adjacent myofibrils can no longer interact coherently, leading to a decrease in the SHG signal [97].

A study on SHG signal efficiency from skin and tendon samples of Sprague Dawley rats during treatment with 100%-glycerol has reported a signal degradation in the course of clearing. After treating the samples, they were rehydrated by application of saline, and the signal returned to its initial state [99]. Since the SHG signal degrades during treatment and recovers after rehydration, the authors of Ref. [99] related the signal variations to the reversible dissociation of collagen fibers and corresponding loss of fibril organization at glycerol action. In this study, the spectral transmittance of the samples was measured between 400 and 700 nm in three states: before treatment, after treatment, and after rehydration. It was observed that after the treatment, transmittance increases significantly and returns to pre-treatment levels after rehydration [99]. Considering the transmittance increase with the treatment, the association of the SHG signal decrease to collagen dissociation is contradictory, since less organized collagen fibers would lead to smaller transmittance [19, 100]. Due to the use of 100%-glycerol in this treatment (no water in the treating solution), the solution provides strong osmotic pressure over the samples, leading to a main effect of tissue dehydration. This way, a more adequate explanation for the SHG signal degradation should be presented. In a reflective-type SHG polarimetry study on chicken skin dermis submitted to dehydration, it was observed that the polarization signal was almost unchanged and the SHG intensity decreased at about a fourth [101]. It was hypothesized by the authors of this study that the decrease of the SHG intensity results from a change of scattering efficiency, rather than a change of the SHG radiation efficiency in tissues. It was also observed that tissue fixation also promotes small changes in SHG polarization, while SHG intensity slightly increases [19, 101]. It is known that formalin fixing induces collagen cross-linking in tissues, suggesting that the cross-linking does not affect collagen orientation but contributes essentially to the efficiency of the SHG signal [101]. From these results, we see that the intensity of the SHG signal depends on light scattering within the sample, which has decreased with tissue dehydration and increased at tissue fixation. This way, in SHG studies of tissue structure (collagen orientation), the optical immersion technique or the SHG polarimetry should be used [19].

   More recent studies have used the optical clearing technique to improve imaging depth and contrast in SHG images from different tissues. One of those studies was reported in 2006 for the case of bovine skeletal muscle [102], where a 2.5-fold increase in SHG imaging depth was obtained with a treatment with 50%-glycerol. After applying signal processing techniques to the acquired signals, it was possible to verify that the applied treatment does not change the periodicity of the sarcomere structure, but it improves image quality deeper in the tissue. Due to the observed lack of glycerol concentration deep in the tissue, the RI matching mechanism was assumed to play a minor role in muscle clearing. Identical clearing treatment was also applied to mouse tendon in this study. A strong decrease in SHG response was observed in this case, which was attributed to the fact that tendon tissue is mainly composed by collagen, in opposition to skeletal muscle that contains mainly myosin [102]. Another study where optical clearing was applied to skeletal muscle tissues from mice has also reported the preservation of myosin heavy chain organization [103]. However, since the samples were previously fixed, the SHG signal from the cleared fibers was considerably weaker than that of the control fibers.

   The combination of SHG with other microscopy imaging techniques has proven beneficial to obtain high-contrast 3D images of tissues in cooperation with optical clearing. One of these studies was recently reported regarding the combination of multiphoton microscopy with OC optimized for pathology evaluation in the kidneys, where the image quality was comparable to traditional histology [104]. Using benzyl alcohol/benzyl benzoate with 4′,6-diamidino-2-phenylindole and eosin in an optimized imaging setup, optical sections with same quality as traditional histology cut-sections were obtained. These images maintained diagnostic-level quality through 1 mm of tissue, and 3D perspectives reveal changes of glomerular capsule cells, which are not seen on single sections. In this study, the SHG signal was measured on transmission, and the data obtained with this technique has allowed for characterizing the collagen matrix organization throughout the entire kidney specimen. The combination of multiphoton sections with SHG data allowed for the 3D image reconstruction with a high contrast [104].

   A similar study was performed on biopsied placental membrane tissues with 6 mm thickness. For control, some of the samples were fixed in 10%-formalin, while the others were treated with 2,2′-thiodiethanol to increase transparency and allow for deeper tissue imaging. Using a multiphoton fluorescence microscope, cellular autofluorescence images were acquired from native and cleared samples. Using same procedures, but with an SHG microscope, images of the fibrillar collagen were also acquired. Optical clearing allowed to perform full-depth imaging from the samples, and those images revealed fetal membrane epithelial topography and collagen organization in multiple matrix layers. Improved visualization of collagen structure, mesenchymal cells in extracellular matrix layers, and decidua morphology within large areas of the biopsied tissues were possible due to optical clearing, and some structural changes in placental tissues due to gestation were identified [105].

   Although SHG imaging at OC action presents certain limitations due to collagen dissociation, some recent studies, where SHG is combined with other imaging techniques, seem to take advantage of such fact to improve in tissue imaging. One of those studies used both multiphoton tomography (MPT) and SHG imaging techniques to probe tissue depth on porcine skin at optical clearing with glycerol and iohexol (Omnipaque™) in different concentrations [12]. After 60 min of topical application of glycerol or Omnipaque™ to the skin samples, a significant improvement on the probing depth and contrast of MPT signals was observed. It was also demonstrated that treatments with different concentrations of glycerol (40%, 60%, and 100%) and Omnipaque™ (60% and 100%) provide improved autofluorescence and SHG signals from the skin. Comparison between results showed that at the applied concentrations and time of treatment, glycerol is more efficient than Omnipaque™ but high concentrations of glycerol induce tissue shrinkage and variations in cell morphology. The authors of this study also verified that the SHG signal intensity increases considerably and it can be detected from the deep skin layers, allowing to obtain information regarding the collagen structure at improved contrast. Comparing between tissues treated with glycerol and Omnipaque™, glycerol is more efficient for SHG signal improvement, even at the lower concentrations [12].

A similar study that also used a combination of MPT and SHG imaging was performed on mice lung to detect excessive collagen deposition, which is indicative of chronic lung diseases. In this study [106], some mice were submitted to nasal administration of polyinosinic:polycytidylic acid (usually abbreviated as poly(I:C)) for 15 days, to induce lung collagen deposition. Twelve days after the last treatment, the animals were sacrificed for lung specimens' collection. An optimized clearing protocol to avoid significant lung shrinkage was adopted in this study. Basically, to dehydrate the lungs, increasing concentrations of methanol in PBS, from 10% (v/v) to 100% at each 10%, were infused into the lung via the cannulated trachea. After dehydration, the RI matching to clear the lungs was made by infusing a solution of benzyl alcohol-benzyl benzoate (1:2) to the lung in the same manner as methanol. After OC treatment, the acquisition of MPT/SHG images was made to obtain a 3D reconstruction of the whole organ. Quantitative analysis of these images revealed significant differences in collagen deposition between the control lungs and the ones treated with poly(I:C). Airway analysis in whole lung showed significant subepithelial fibrosis between the proximal and distal airways for the group treated with poly(I:C) [106].

With the objective of creating a skull window for assessing cerebrovascular structure and function, Zhu's group has used two optical clearing solutions to treat the skulls of mice [107]. The first solution consists on uniform mixture of 75% (v/v) of ethanol with 25% of urea, while the second consists on a mixture of 0.7 M NaOH and dodecylbenzenesulfonic acid at a volume-mass ratio of 24:5. These solutions were topically applied to mice skulls to create a skull window within 15 min, and enhancement both in contrast and imaging depth was observed. Apart from being able to see cerebrovascular structure and function through the created window, the evaluation of the SHG signal was also performed. A 10-min treatment with the first clearing solution showed evident reduction of the SHG signal intensity and revealed the occurrence of some cavities at the skull surface as a result of collagen dissolution. For a 5-min treatment with the second clearing solution, the SHG signal intensity seemed unchanged. For both treatments and after recovery with PBS scrubbing, the skull SHG signals have increased, and cavities were reduced, indicating the reassembly of collagen structures [107].

The ability of cooperating tissue OC with SHG imaging to detect rich collagen concentrations in soft and hard tissues was also recently reported. A study was conducted on the entire mouse body after treatment with polyethylene glycol (PEG)-associated solvent system (PEGASOS) [108]. This clearing method does not eliminate autofluorescence, and after the treatment nearly all the types of tissues in the mouse body became transparent. Through a combination of SHG with two-photon or fluorescence microscopy, it was possible to acquire images of independent organs inside the mouse body with optimized resolution. The SHG technique allowed to detect large concentrations of collagen in tissues such as bones or teeth. 3D imaging of the mouse head with teeth, brain, and muscle discrimination without blind spots was possible. 3D reconstruction of the mandible, teeth, femur, and knee joint was obtained from the fluorescence images, and subcellular resolution of the intact mouse brain and neuron/axon tracing over a long distance was made possible. A piece of dog tibia was cleared in the same manner as a large hard tissue, and the SHG signal was measured from 3 mm depth [108].

7.7 The Light-Sheet Method

An imaging technique that has undergone revolutionary evolution in the last decade is the light-sheet microscopy. This method consists of illuminating a biological specimen with a thin light sheet through a normal direction to the observing axis of the microscope to obtain optical sectioning without physical sectioning. Since excitation and detection are organized in orthogonal directions and a single sweep without line scanning acquires a single x–y plane image, this technique enables high volumetric imaging speed with high signal-to-noise ratio and low levels of photobleaching and phototoxic effects [109–113].

This imaging technique has a long history, since its discovery in 1912 by a German chemist, Richard Zsigmondy, and a German physicist, Henry Siedentopf. These researchers wanted to improve the resolution of optical microscopes to observe nanostructured materials such as colloidal gold [114]. By that time, and as a result of the work by Abbe, the resolution improvement in microscopy was limited, and to reach it, it was necessary to use lenses with high numerical aperture (NA) and short wavelengths of light. In benefit of Zsigmondy and Siedentopf, an employee of Zeiss™ at that time realized that to maximize visibility of any features at the microscope, in addition to improved resolution, high contrast was also necessary. By illuminating the sample through a perpendicular direction to the observation axis of the microscope, they could observe only the light that was scattered by the gold nanoparticles into their line of observation. In 1912, although powerful, the light introduced in the microscope was merely sunlight that was channeled for the purpose. On the other hand, since no microscope photography was available at that time, the researcher needed to observe with his/her own eye. Regardless of these instrumental limitations, it was possible to see different colors for particles with different sizes and from different materials [114].

During the second half of the twentieth century, with the development of microscopic cameras and portable semiconductor lasers, it was possible to use the light-sheet microscopy to analyze small volumes within tissues or organs. The first application of the light-sheet scheme for fluorescence microscopy was published in 1993, where high-resolution images of the internal structure of the guinea pig cochlea were published [115]. In this study, the lateral and axial resolutions obtained in the acquired images were 10 μm and 26 μm, respectively, within a 1–5 mm field of view.

Further development and improvement of selective plane illumination microscopy (SPIM) started in 2004 with an imaging study of embryos from the fruit fly (*Drosophila melanogaster*) [116]. Since then, this imaging technique has been upgraded constantly and widely applied to various situations. The first ultramicroscope with fluorescence excitation and limited resolution was made available commercially in 2010 [117].

A typical scheme for the light-sheet microscope is presented in Fig. 7.5, where sample illumination is made from one side only.

In the scheme presented in Fig. 7.5, a laser beam passes through a beam expander to increase the beam diameter, so that it can fill the cylindrical lens that will produce the light sheet. Then the light sheet is directed to the specimen through an illuminating objective to reduce light aberrations produced by the cylindrical lens. Two lasers (532 and 488 nm) are commonly used with this setup to visualize a wide range of biologically useful fluorophores, such as rhodamine, fluorescein, and fluorescent proteins [118]. The scattered light at the specimen is collected by the detecting

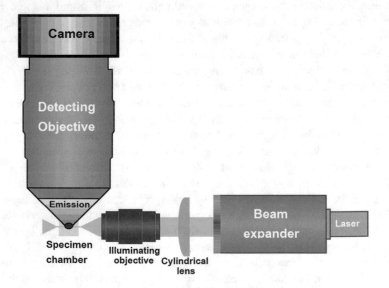

Fig. 7.5 Light-sheet microscope scheme (adapted from Ref. [118])

objective of the microscope for visual observation or for photographic record with the camera on the top. The illumination branch has translational movement to allow collecting several sheet images that with appropriate software can be "stacked" to create a 3D image of the specimen.

Some other variations to the scheme presented in Fig. 7.5 use a second illuminating branch, placed in opposition to the first one. This way, double-sided illumination of the specimen chamber reduces shadow artifacts, provides a more intense beam, and allows illumination of a larger specimen. The dual-sided illumination presents a significant disadvantage, since it is difficult to align perfectly both light sheets in three-dimensional space, resulting into a non-well focused optical section as in the single-sided illumination [118].

The combination of this technique with OC has emerged naturally to improve contrast and resolution and also to image deeper tissue layers from a thick tissue. One of the first studies where OC was applied with light-sheet microscopy was performed to image the neural pathways in rats [119]. In this study, which was reported in 2012, a confocal light-sheet microscope (CLSM) was used to obtain section images of the rat brain after clearing and reconstruct large volumes with subcellular resolution. The whole brain or excised hippocampus and cerebella specimens were first submitted to sequential dehydration procedures in increasing concentrations of ethanol solutions from ~30% to ~100%. The immersion in each solution took 2 h, and the final dehydration stage in 100%-ethanol was performed overnight. To perform the RI matching, the dehydrated excised specimens were treated with 1:2 benzyl alcohol and benzyl benzoate for 36 h. The whole brain specimens were treated with same solution for ~3 days. All dehydration and clearing procedures were performed at 18–22 °C. A comprehensive map of the Purkinje cells in the cerebellum was obtained, and a trace of neuronal projections across the brain was created [119].

The combination of light-sheet microscopy with OC has also allowed to image rat mammary tissues. After treating a normal mammary tissue with clear unobstructed brain imaging cocktails (CUBIC), a $0.8 \times 0.8 \times 1.5$ mm volume could be imaged with dual-sided illumination in less than 5 min [120]. To compare imaging results from different techniques, Lloyd-Lewis et al. used a two-photon microscope to image a mammary tumor after clearing with See Deep Brain protocol (SeeDB). As a result, the imaging of a $1.2 \times 1.2 \times 0.1$ mm volume was obtained at high cellular resolution after 15 min [120].

A particular study of the physiological processes in bone tissue, such as hematopoiesis, bone growth and bone remodeling, was performed also using the light-sheet technique and OC treatments [121]. Greenbaum et al. developed a bone tissue clearing protocol (designated as Bone CLARITY) and used it with a custom-built light-sheet fluorescence microscope to detect the endogenous fluorescence of Sox9-tdTomato osteoprogenitor cells in tibia, femur, and vertebral column of adult transgenic mice. The OC protocol takes several days to clear the bones and uses an acrylamide hydrogel to support the tissue structure and minimize protein loss before the delipidation step. Sodium dodecyl sulfate (SDS) is used to remove lipids, so that their light-scattering effects can be minimized. High autofluorescence was observed in one of the primary sites of heme synthesis—the bone marrow. To reduce this excessive autofluorescence by about threefold, the heme was removed with amino alcohol N,N,N',N'-tetrakis (2-hydroxypropyl) ethylenediamine. All these OC procedures were conducted at controlled temperature, and no size changes were observed. The whole bone was cleared, and images could be obtained from as deep as 1.5 mm, with unchanged endogenous fluorescence and a signal-to-noise ratio that allowed image detection and 3D reconstruction of single cells [121].

The light-sheet imaging technique has also been applied with OC to obtain images of the vasculature in retinal tissues from pup rats [122]. In this study, after tissue preparation and protein labeling, the OC was made using RI matched solution (RIMS), a solution containing 40 g of Histodenz in 30 mL of 0.02 M phosphate buffer with 0.1% of Triton X-100. When imaging the retina samples at the light-sheet microscope, the control samples showed dense and consistent vascular distributions with narrow avascular zones at the retinal periphery, while flat retina mounts that suffered prolonged hyperoxia (HO) exposure showed simplified vessel branching and decreased vessel concentration throughout the primary plexus, with large avascular zones at the retinal periphery. When imaging retinas from within the untouched eyes, the nature structure of the eye was seen to be better preserved, and volumetric imaging was performed fast, at ~100 to 500 ms per image plane. The presence of the choroid layer of vasculature was detected in individual image planes within the volumetric images, and the primary and secondary plexus were found at the deepest portion of the retina. It was also observed that retinal vasculature was more developed in eyes from rats that were sacrificed 14 days after birth than in rats that were sacrificed 7 days after birth [122].

Using a different OC protocol, Henning et al. have also studied eye and retinal vasculature from mouse intact eyeballs [123]. In this protocol, the eyeball clearing consisted on ethanol dehydration followed by RI matching, which was provided mainly by ethyl cinnamate (ECi). One of the treatment steps of this protocol consists of a highly effective melanin bleaching procedure. Such protocol is compatible with immunolabeling to allow ocular and retinal vasculature visualization with a light-sheet fluorescence microscopy. One drawback observed in this study was that retina detaches from the retinal pigment epithelium (RPE), possibly caused by tissue shrinkage during dehydration. Regardless of this disadvantage, the acquired images allowed one to perform a high-resolution analysis of vascular architecture in the entire eyeballs. Studies of eye vasculature can provide means of discriminating eye diseases, such as retinal degenerative diseases [123].

Neural networks in the brain, brain vasculature, and its network are also of interest to study, since cerebral health can be monitored through the visualization of neural circuits and vascular delivery of oxygen and nutrients. In such cases, it is interesting to have a high contrast and resolution combined with a wide-volume 3D perspective of the brain. The combination of OC treatments and light-sheet imaging is perfect to produce such brain images. A few studies on brain vasculature and neural circuits have been made recently.

One of those studies used a novel approach that consisted of filling the entire blood vessel with a fluorescent gel for improved contrast imaging

[124]. Combining this approach with OC treatment, imaging of the brain was made using a light-sheet fluorescence microscopy at improved contrast, in particular at higher depth. After collecting and preparing mice brains, they were cleared with 50 mL of 30% 2,2'-thiodiethanol (TDE)-PBS for 1 day, followed by 2 days of incubation in 63% TDE/PBS, both at 37 °C in gentle shaking. A custom-made light-sheet microscope, with fluorescent excitation at 491 and 561 nm, was used to image the entire brains. A custom-made chamber filled with the mounting medium (63% TDE/PBS) contained a quartz cuvette, where the brain was placed for imaging. Planar images of the brain layers were acquired with a field of view of 1.3 × 1.3 mm and a xy resolution of 0.65 μm. The z-step between image layers was 2 μm. With the high resolution of the light-sheet imaging method over a cleared brain and the contrast-enhanced vessels, it was possible to map both the neural and vascular circuits in the entire mouse brain. This technique produces high-quality results, and it can be used to detect physiological alterations of the brain connected to some diseases, such as Alzheimer's condition [124].

Pavone's group developed another study to compare the image results obtained at the light-sheet microscope, when excited with Gaussian and Bessel beam types. Using intact optically cleared mouse brains, they obtained that Bessel beams produce high-fidelity structural data for the wide-brain morphology of neuronal and vascular networks with reduced streaking artifacts. Gaussian beams, on the other hand, produce a significant loss of information in the acquired images, and intensity inhomogeneity affects about a third of the imaged volume [125].

Following the CUBIC and a modified IDISCO (iDISCO+) clearing protocols (see Refs. [5] and [126] for more details), Rocha et al. have used a light-sheet microscopy to image the entire brain of the *Zebra Finch* bird at cellular resolution [127]. Although both clearing protocols created acceptable transparency in the bird's brain, authors found that prolonging the delipidation step in the CUBIC protocol helps to obtain better transparency. In the case of the CUBIC protocol, vasculature of the forebrain was clearly seen in images of the autofluorescence structures, while anatomical landmarks were detected from the tissue treated with the iDISCO+ protocol [127].

An alternative to the tissue immersion OC in light-sheet imaging of mouse brain tissues was recently reported [128]. To improve contrast and resolution, Bürgers et al. have expanded the tissue samples. In this method, water adsorbent polymers are used to physically expand enzymatically treated tissue samples in an isotropic manner, resulting in the fact that fluorescent moieties that have a closer distribution than the optical diffraction limit (≈250 nm) can be optically resolved [128]. Similarly to the immersion OC procedure, the expansion method intrinsically turns the tissues transparent due to their high water content and enlarged distance between scatterers and absorbers. Using this method, designated as the light-sheet fluorescence expansion microscopy (LSFEM), allowed the rapid acquisition of super-resolved neuronal connectivity maps from mouse hippocampal subregions using a diffraction limited light-sheet microscope. Bürgers et al. could image and segment individual granule cells inside the densely packed granule cell layer from the dorsal hippocampus, together with their neuritis extending deep into the molecular layer in super-resolution revealing their fine structural details. According to authors of Ref. [128], the imaging resolution was high enough to visualize individual dendritic spines and acquired data allowed to demonstrate two-color labeling of pre- and postsynaptic proteins.

References

1. M. Li, *Developing a Technique for Combining Light and Ultrasound for Deep Tissue Imaging* (MS thesis), Sweden: Lund University, 2018
2. E.A. Genina, A.N. Bashkatov, V.V. Tuchin, Tissue optical immersion clearing. Expert Rev. Med. Devices **7**(6), 825–842 (2010)
3. V.V. Tuchin, *Optical Clearing of Tissues and Blood* (SPIE Press, Bellingham, WA, 2006)
4. L. Oliveira, M.I. Carvalho, E.M. Nogueira, V.V. Tuchin, Diffusion characteristics of ethylene glycol in skeletal muscle. J. Biomed. Opt. **20**(5), 051019 (2015)
5. D.S. Richardson, J.W. Lichtman, Clarifying tissue clearing. Cell **162**, 246–257 (2015)
6. P.S. Tsai, P. Blinder, B.J. Migliori, J. Neev, Y. Jin, J.A. Squier, D. Kleinfeld, Plasma-mediated ablation: an optical tool for submicrometer surgery on neuronal and vascular systems. Curr. Opin. Biotech. **20**, 90–99 (2009)
7. S. Carvalho, N. Gueiral, E. Nogueira, R. Henrique, L. Oliveira, V.V. Tuchin, Comparative study of the optical properties of colon mucosa and colon precancerous polyps between 400 and 1000 nm, in Dynamics and Fluctuations in Biomedical Photonics XIV, ed. by V.V. Tuchin, K.V. Larin, M.J. Leahy, R.K. Wang. Proc. SPIE **10063**, 100631L (2017)
8. H. Duong, M. Han, A multispectral LED array for the reduction of background autofluorescence in brain tissue. J. Neurosci. Methods **220**, 46–54 (2013)
9. B. Clancy, L.J. Cauller, Reduction of background autofluorescence in brain sections following immersion in sodium borohydride. J. Neurosci. Methods **83**, 97–102 (1998)
10. T. Zimmermann, Spectral imaging and linear unmixing in light microscopy. Adv. Biochem. Eng. Biotechnol. **95**, 245–265 (2005)
11. A.Y. Sdobnov, V.V. Tuchin, J. Lademann, M.E. Darvin, Confocal Raman microscopy supported by optical clearing treatment of the skin – influence on collagen hydration. J. Phys. D Appl. Phys. **50**(28), 285401 (2017)
12. A.Y. Sdobnov, M.E. Darvin, J. Lademann, V.V. Tuchin, A comparative study of ex vivo skin optical clearing using two-photon microscopy. J. Biophotonics **10**(9), 1115–1123 (2017)
13. A.Y. Sdobnov, M.E. Darvin, J. Schleusener, J. Lademann, V.V. Tuchin, Hydrogen bound water profiles in the skin influenced by optical clearing molecular agents-quantitative analysis using confocal Raman microscopy. J. Biophotonics **12**(5), e201800283 (2019)
14. A.Y. Sdobnov, M.E. Darvin, E.A. Genina, A.N. Bashkatov, J. Lademan, V.V. Tuchin, Recent progress in tissue optical clearing for spectroscopic application. Spectrochim. Acta Part A: Mol. Biomol. Spectrosc. **197**, 216–229 (2018)
15. A.Y. Sdobnov, J. Lademann, M.E. Darvin, V.V. Tuchin, Methods for optical skin clearing in molecular optical imaging in dermatology. Biochem. **84**(S1), 144–158 (2019)
16. D. Huang, E.A. Swanson, C.P. Lin, J.S. Schuman, W.G. Stinson, W. Chang, M.R. Hee, T. Flotte, K. Gregory, C.A. Puliafito, J.G. Fujimoto, Optical coherence tomography. Science **254**, 1178–1181 (1991)
17. A.F. Fercher, K. Mengedoht, W. Werner, Eye-length measurement by interferometry with partially coherent light. Opt. Lett. **13**(3), 186–188 (1988)
18. W. Drexler, J. G. Fujimoto (eds.), *Optical Coherence Tomography: Technology and Applications*, 2nd edn. (Springer International Publishing Switzerland, Cham, 2015)
19. V.V. Tuchin, *Tissue Optics – Light Scattering Methods and Instruments for Medical Diagnostics*, 3rd edn. (SPIE Press, Bellingham, WA, 2015)

20. https://en.wikipedia.org/wiki/Optical_coherence_tomography. Accessed 25 Jul 2019
21. A.F. Fercher, Optical coherence tomography. J. Biomed. Opt. **1**, 157–173 (1996)
22. E.A. Genina, A.N. Bashkatov, M.D. Kozintseva, V.V. Tuchin, OCT study of optical clearing of muscle tissue in vitro with 40% glucose solution. Opt. Spect. **120**(1), 27–35 (2016)
23. Y.M. Liew, R.A. McLaughlin, F.M. Wood, D.D. Sampson, Reduction of image artifacts in three-dimensional optical coherence tomography of skin in vivo. J. Biomed. Opt. **16**(11), 116018 (2011)
24. E.A. Genina, A.N. Bashkatov, Y.P. Sinichkin, I.Y. Yanina, V.V. Tuchin, Optical clearing of biological tissues: prospects of application in medical diagnostics and phototherapy. J. Biomed. Phot. Eng. **1**(1), 22–58 (2015)
25. R.K. Wang, V.V. Tuchin, Optical coherence tomography. Light scattering and imaging enhancement, Chapter 16, in *Handbook of Coherent-Domain Optical Methods: Biomedical Diagnostics, Environmental Monitoring, and Material Science*, ed. by V. V. Tuchin, vol. 2, 2nd edn., (Springer, New York, NY, 2013), p. 665
26. A.N. Bashkatov, E.A. Genina, V.I. Kochubey, V.V. Tuchin, Optical properties of human sclera in spectral range 370-2500 nm. Opt. Spectr. **109**(2), 197–204 (2010)
27. A.N. Bashkatov, E.A. Genina, V.I. Kochubey, V.V. Tuchin, Optical properties of human skin, subcutaneous and mucous tissues in the wavelength range from 400 to 2000 nm. J. Phys. D Appl. Phys. **38**(15), 2543–2555 (2005)
28. A.N. Bashkatov, E.A. Genina, M.D. Kozintseva, V.I. Kochubey, S.Y. Gorofkov, V.V. Tuchin, Optical properties of peritoneal biological tissues in the spectral range of 350-2500 nm. Opt. Spectr. **120**(1), 1–8 (2016)
29. I. Carneiro, S. Carvalho, R. Henrique, L.M. Oliveira, V.V. Tuchin, Optical properties of colorectal muscle in visible/NIR range, in Biophotonics: Photonic Solutions for Better Health Care VI, ed. by J. Popp, V.V. Tuchin, F.S. Pavone. Proc. SPIE **10685**, 106853D (2018)
30. A.N. Bashkatov, E.A. Genina, V.V. Tuchin, Optical properties of skin, subcutaneous and muscle tissues: a review. J. Innov. Opt. Health Sci. **4**(1), 9–38 (2011)
31. M.G. Ghosn, V.V. Tuchin, K.V. Larin, Nondestructive quantification of analyte diffusion in cornea and sclera using optical coherence tomography. Invest. Ophthalmol. Vis. Sci. **48**(6), 2726–2733 (2007)
32. A.N. Bashkatov, E.A. Genina, V.V. Tuchin, Measurement of glucose diffusion coefficients in human tissues, Chapter 19, in *Handbook of Optical Sensing of Glucose in Biological Fluids and Tissues*, ed. by V. V. Tuchin, (Taylor & Francis Group LLC, CRC Press, Boca Raton, FL, 2009), pp. 587–621
33. M.G. Ghosn, E.F. Carbajal, N.A. Befrui, V.V. Tuchin, K.V. Larin, Differential permeability rate and percent clearing of glucose in different regions in rabbit sclera. J. Biomed. Opt. **13**(2), 021110 (2008)
34. X. Guo, G. Wu, H. Wei, X. Deng, H. Yang, Y. Ji, Y. He, Z. Guo, S. Xie, H. Zhong, Q. Zhao, Z. Zhu, Quantification of glucose diffusion in human lung tissues by using Fourier domain optical coherence tomography. Potochem. Photobiol. **88**, 311–316 (2012)
35. K.V. Larin, M.G. Ghosn, A.N. Bashkatov, E.A. Genina, N.A. Trunina, V.V. Tuchin, Optical clearing for OCT image enhancement and in-depth monitoring of molecular diffusion. IEEE J. Sel. Top. Quant. Elect. **18**(3), 1244–1259 (2012)
36. R. Wang, V.V. Tuchin, Enhance light penetration in tissue for high resolution optical imaging techniques by the use of biocompatible chemical agents. J. X ray Sci. Tech. **10**, 167–176 (2002)
37. R.K. Wang, J.B. Elder, Propylene glycol as a contrasting agent for optical coherence tomography to image gastrointestinal tissues. Lasers Surg. Med. **30**, 201–208 (2002)
38. L. Pires, V. Demidov, I.A. Vitkin, V. Bagnato, C. Kurachi, B.C. Wilson, Optical clearing of melanoma in vivo: characterization by diffuse reflectance spectroscopy and optical coherence tomography. J. Biomed. Opt. **21**(8), 081210 (2016)

39. Z. Zhu, G. Wu, H. Wei, H. Yang, Y. He, S. Xie, Q. Zhao, X. Guo, Investigation of the permeability and optical clearing ability of different analytes in human normal and cancerous breast tissues by spectral domain OCT. J. Biophotonics **5**(7), 536–543 (2012)

40. H. Xiong, Z. Guo, C. Zeng, L. Wang, Y. He, S. Liu, Application off hyperosmotic agent to determine gastric cancer with optical coherence tomography ex vivo in mice. J. Biomed. Opt. **14**(2), 024029 (2009)

41. H.Q. Zhong, Z.Y. Guo, H.J. Wei, J.L. Si, L. Guo, Q.L. Zhao, C.C. Zheng, H.L. Xiong, Y.H. He, S.H. Liu, Enhancement of permeability of glycerol with ultrasound in human normal and cancer breast tissues in vitro using optical coherence tomography. Laser Phys. Lett. **7**(5), 388–395 (2010)

42. E.A. Genina, A.N. Bashkatov, O.V. Semyachkina-Glushkovskaya, V.V. Tuchin, Optical clearing of cranial bone by multicomponent immersion solutions and cerebral venous blood flow visualization. Izv. Saratov Univ. (N. S.), Ser. Phys. **17**, 98–110 (2017)

43. H. Zhong, Z. Guo, H. Wei, L. Uo, C. Wang, Y. He, H. Xiong, S. Liu, Synergistic effect of ultrasound and thiazone-PEG 400 on human skin optical clearing in vivo. Photochem. Photobiol. **86**, 732–737 (2010)

44. E.A. Genina, A.N. Bashkatov, E.A. Kolesnikova, M.V. Basko, G.S. Terentyuk, V.V. Tuchin, Optical coherence tomography monitoring of enhanced skin optical clearing in rats in vivo. J. Biomed. Opt. **19**(2), 021109 (2014)

45. W. Feng, R. Shi, N. Ma, D.K. Tuchina, V.V. Tuchin, D. Zhu, Skin optical clearing potential of disaccharides. J. Biomed. Opt. **21**(8), 081207 (2016)

46. L. Guo, R. Shi, C. Zhang, D. Zhu, Z. Ding, P. Li, Optical coherence tomography angiography offers comprehensive evaluation of skin optical clearing in vivo by quantifying optical properties and blood flow imaging simultaneously. J. Biomed. Opt. **21**(8), 081202 (2016)

47. N.S. Ksenofontova, E.A. Genina, A.N. Bashkatov, G.S. Terentyuk, V.V. Tuchin, OCT study of skin optical clearing with preliminary laser ablation of epidermis. J. Biomed. Phot. Eng. **3** (2), 020307 (2017)

48. O. Zhernovaya, V.V. Tuchin, M.J. Leahy, Enhancement of OCT imaging by blood optical clearing in vessels – a feasibility study. Photon. Lasers Med. **5**(2), 151–159 (2016)

49. A. Bykov, T. Hautala, M. Kinnunen, A. Popov, S. Karhula, S. Saarakkala, M.T. Nieminen, V.V. Tuchin, I. Meglinski, Imaging of subchondral bone by optical coherence tomography upon optical clearing of articular cartilage. J. Biophotonics **9**(3), 270–275 (2016)

50. C.H. Liu, M. Singh, J. Li, Z. Han, C. Wu, S. Wang, R. Idugboe, R. Raghunathan, E.N. Sobol, V.V. Tuchin, M. Twa, K.V. Larin, Quantitative assessment of hyaline cartilage elasticity during optical clearing using optical coherence elastography. Med. Technol. Med **7**, 44–51 (2015)

51. A.F. Peña, A. Doronin, V.V. Tuchin, I. Meglinski, Monitoring of interaction of low-frequency electric field with biological tissues upon optical clearing with optical coherence tomography. J. Biomed. Opt. **19**(8), 086002 (2014)

52. https://en.wikipedia.org/wiki/Fluorescence_microscope. Accessed 25 Jul 2019

53. I.V. Meglinski, A.N. Bashkatov, E.A. Genina, D.Y. Churmakov, V.V. Tuchin, The enhancement of confocal images of tissues at bulk optical immersion. Laser Phys. **13**(1), 65–69 (2003)

54. I.V. Meglinski, A.N. Bashkatov, E.A. Genina, D.Y. Churmakov, V.V. Tuchin, Study of the possibility of increasing the probing depth by the method of reflection confocal microscopy upon immersion clearing of near-surface human skin layers. Quant. Elect. **32**(10), 875–882 (2002)

55. R. Dickie, R.M. Bachoo, M.A. Rupnick, S.M. Dallabrida, G.M. Deloid, J. Lai, R.A. DePinho, R.A. Rogers, Three-dimensional visualization of microvessel architecture of whole-mount tissue by confocal microscopy. Microvasc. Res. **72**, 20–26 (2006)

56. A.-S. Chiang, Y.-C. Liu, S.-L. Chiu, S.-H. Hu, C.-Y. Huang, C.-H. Hsieh, Three-dimensional mapping of brain neuropils in the cockroach Diploptera punctate. J. Comp. Neurol. **440**, 1–11 (2001)

57. A.J. Moy, B.V. Capulong, R.B. Saager, M.P. Wiersma, P.C. Lo, A.J. Durkin, B. Choi, Optical properties of mouse brain tissue after optical clearing with FocusClear. J. Biomed. Opt. **20**(9), 095010 (2015)

58. Y.-Y. Fu, C.-W. Lin, G. Enikolopov, E. Sibley, A.-S. Chiang, S.-C. Tang, Microtome-free 3-dimensional confocal imaging method for visualization of mouse intestine with subcellular-level resolution. Gastroenterology **137**, 453–465 (2009)

59. Y. Aoyagi, R. Kawakami, H. Osanai, T. Hibi, T. Nemoto, A rapid optical clearing protocol using 2,2′-thiodiethanol for microscopic observation of fixed mouse brain. PLoS One **10**(1), e0116280 (2015)

60. I. Costantini, J.-P. Ghobril, A.P. Di Giovanna, A.L.A. Mascaro, L. Silvestri, M.C. Müllenbroich, L. Onofri, V. Conti, F. Vanzi, L. Sacconi, R. Guerrini, H. Markram, G. Lannelo, F.S. Pavone, A versatile clearing agent for multi-modal brain imaging. Sci. Rep. **5**, 9808 (2015)

61. J. Hasegawa, Y. Sakamoto, S. Nakagami, M. Aida, S. Sawa, S. Matsunaga, Three-dimensional imaging of plant organs using a simple and rapid transparency technique. Plant Cell Physiol. **57**(3), 462–472 (2016)

62. T.J. Musielak, D. Slane, C. Liebig, M. Bayer, A versatile optical clearing protocol for deep tissue imaging of fluorescent proteins in Arabidopsis thaliana. PLoS One **11**(8), e0161107 (2016)

63. W. Li, R.N. Germain, M.Y. Gerner, Multiplex, quantitative cellular analysis in large tissue volumes with clearing-enhanced 3D microscopy (C_e3D). Proc. Natl. Sci. Acad. U S A **114** (35), E7321–E7330 (2017)

64. Online supporting information appendix to reference 64. https://www.pnas.org/content/pnas/suppl/2017/08/08/1708981114.DCSupplemental/pnas.1708981114.sapp.pdf. Accessed 5 Apr 2019

65. R. Samatham, K.G. Phillips, S.L. Jacques, Assessment of optical clearing agents using reflectance-mode confocal scanning laser microscopy. J. Innov. Opt. Health Sci. **3**, 183–188 (2010)

66. C.P. Neu, T. Novak, K.F. Gilliland, P. Marshall, S. Calve, Optical clearing in collagen- and proteoglycan-rich osteochondral tissues. Osteoarthr. Cartil. **23**, 405–413 (2015)

67. M. Pende, K. Becker, M. Wanis, S. Saghafi, R. Kaur, C. Hahn, N. Pende, M. Foroughipour, T. Hummel, H.-U. Dodt, High-resolution ultramicroscopy of the developing and adult nervous system in optically cleared Drosophila melanogaster. Nat. Commun. **9**, 4731 (2018)

68. H. Hama, H. Hioki, K. Namiki, T. Hoshida, H. Kurokawa, F. Ishidate, T. Kaneko, T. Akagi, T. Saito, T. Saido, A. Miyawaki, ScaleS: an optical clearing palette for biological imaging. Nat. Neurosci. **18**(10), 1518–1529 (2015)

69. L. Liu, A. Liu, W. Xiao, R. Li, X. Hu, L. Chen, Volumetric fluorescence imaging combined with modified optical clearing Alzheimer's disease pathology, in *2018 Asia Communications and Photonics Conference (ACP), Hangzhou*, (IEEE, New York, NY, 2018), pp. 1–3. https://doi.org/10.1109/ACP.2018.8596080

70. A.F. Fercher, J.D. Briers, Flow visualization by means of single-exposure speckle photography. Opt. Commun. **37**(5), 326–330 (1981)

71. J.D. Briers, Laser Doppler, speckle and related techniques for blood perfusion mapping and imaging. Physiol. Meas. **22**, 35–66 (2001)

72. A.K. Dunn, H. Bolay, M.A. Moskowitz, D.A. Boas, Dynamic imaging of cerebral blood flow using laser speckle. J. Cereb. Blood Flow Metab. **21**, 195–201 (2001)

73. A.K. Dunn, Laser speckle contrast imaging of cerebral blood flow. Ann. Biomed. Eng. **40**, 367–377 (2012)

74. D. Briers, D.D. Duncan, E. Hirst, S.J. Kirkpatrick, M. Larsson, W. Steenbergen, T. Stromberg, O.B. Thompson, Laser speckle contrast imaging: theoretical and practical limitations. J. Biomed. Opt. **18**(6), 066018 (2013)

75. I. Sigal, R. Gad, A.M. Caravaca-Aguirre, Y. Atchia, D.B. Conkey, R. Piestun, O. Levi, Laser speckle contrast imaging with extended depth of field for in-vivo tissue imaging. Biomed. Opt. Express **5**(1), 123–135 (2014)
76. A.N. Bashkatov, K.V. Berezin, K.N. Dvoretskiy, M.L. Chernavina, E.A. Genina, V.D. Genin, V.I. Kochubey, E.N. Lazareva, A.B. Pravdin, M.E. Shvachkina, P.A. Timoshina, D.K. Tuchina, D.D. Yakovlev, D.A. Yakovlev, I.Y. Yanina, O.S. Zhernovaya, V.V. Tuchin, Measurement of tissue optical properties in the context of tissue optical clearing. J. Biomed. Opt. **23**(9), 091416 (2018)
77. J. Wang, D. Zhu, Switchable skin window induced by optical clearing method for dermal blood flow imaging. J. Biomed. Opt. **18**(6), 061209 (2013)
78. Q. Luo, C. Jiang, P. Li, H. Cheng, Z. Wang, Z. Wang, V.V. Tuchin, Laser speckle imaging of cerebral blood flow, Chapter 5, in *Coherent-Domain Optical Methods: Biomedical Diagnostics, Environmental Monitoring and Material Science*, ed. by V. V. Tuchin, vol. 1, 2nd edn., (Springer, New York, NY, 2013), pp. 167–212
79. E.I. Galanzha, V.V. Tuchin, A.V. Solovieva, T.V. Stepanova, Q. Luo, H. Cheng, Skin backreflectance and microvascular system functioning at the action of osmotic agents. J. Phys. D Appl. Phys. **36**, 1739–1746 (2003)
80. D. Zhu, J. Zhang, H. Cui, Z. Mao, P. Li, Q. Luo, Short-term and long-term effects of optical clearing agents on blood vessels in chick chorioallantoic membrane. J. Biomed. Opt. **13**(2), 021106 (2008)
81. D. Zhu, J. Wang, Z. Zhi, X. Wen, Q. Luo, Imaging dermal blood flow through the intact rat skin with an optical clearing method. J. Biomed. Opt. **15**(2), 026008 (2010)
82. R. Shi, M. Chen, V.V. Tuchin, D. Zhu, Accessing to arteriovenous blood flow dynamics response using combined laser speckle contrast imaging and skin optical clearing. Biomed. Opt. Express **6**(6), 1977–1989 (2015)
83. Z. Mao, X. Wen, J. Wang, D. Zhu, The biocompatibility of the dermal injection of glycerol in vivo to achieve optical clearing. Proc. SPIE **7519**, 75191N (2009)
84. W. Feng, R. Shi, C. Zhang, S. Liu, T. Yu, D. Zhu, Visualization of skin microvascular dysfunction of type 1 diabetic mice using in vivo skin optical clearing method. J. Biomed. Opt. **24**(3), 031003 (2019)
85. J. Wang, Y. Zhang, T.H. Xu, Q.M. Luo, D. Zhu, An innovative transparent cranial window based on skull optical clearing. Laser Phys. Lett. **9**(6), 469–473 (2012)
86. P.A. Timoshina, A.B. Bucharskaya, D.A. Alexandrov, V.V. Tuchin, Study of blood microcirculation of pancreas in rats with alloxan diabetes by laser speckle contrast imaging. J. Biomed. Phot. Eng. **3**(2), 020301 (2017)
87. P.A. Timoshina, E.M. Zinchenko, D.K. Tuchina, M.M. Sagatova, O.V. Semyachkina-Glushkovskaya, V.V. Tuchin, Laser speckle contrast imaging of cerebral blood flow of newborn mice at optical clearing. Proc. SPIE **10336**, 1033610 (2017)
88. P.A. Timoshina, A.B. Bucharskaya, N.A. Navolokin, V.V. Tuchin, Speckle-contrast imaging of pathological tissue microhemodynamics at optical clearing, in Dynamics and Fluctuations in Biomedical Photonics XVI. Proc. SPIE **10877**, 108770Z (2019). https://doi.org/10.1117/12.2508794
89. D. Abookasis, T. Moshe, Feasibility study of hidden flow imaging based on laser speckle technique using multiperspectives contrast images. Opt. Lasers Eng. **62**, 38–45 (2014)
90. D. Abookasis, T. Moshe, Reconstruction enhancement of hidden objects using multiple speckle contrast projections and optical clearing agents. Opt. Commun. **400**, 58–64 (2013)
91. T. Moshe, M.A. Firer, D. Abookasis, Object reconstruction in scattering medium using multiple elliptical polarized speckle contrast projections and optical clearing agents. Opt. Lasers Eng. **68**, 172–179 (2015)
92. J. Wang, N. Ma, R. Shi, Y. Zhang, T. Yu, D. Zhu, Sugar-induced skin optical clearing: from molecular dynamics simulation to experimental demonstration. IEEE J Sel. Top. Quant. Elect. **20**(2), 7101007 (2014)

93. I. Freund, M. Deutsch, Second-harmonic microscopy of biological tissue. Opt. Lett. **11**(2), 94–96 (1986)

94. P.J. Campagnola, C.Y. Dong, Second harmonic generation microscopy: principles and applications to disease diagnosis. Laser Phot. Rev. **5**(1), 13–26 (2011)

95. P. Campagnola, H.A. Clark, W.A. Mohler, A. Lewis, L.M. Loew, Second-harmonic imaging microscopy of living cells. J. Biomed. Opt. **6**(3), 277–286 (2001)

96. X. Chen, O. Nadiarynkh, S. Plotnikov, P.J. Campagnola, Second harmonic generation microscopy for quantitative analysis of collagen fibrillar structure. Nat. Protoc. **7**(4), 654–669 (2012)

97. O. Nadiarnykh, P.J. Campagnola, SHG and optical clearing, in *Second Harmonic Generation Imaging*, ed. by F. S. Pavone, P. J. Campagnola, (CRC Press, Boca Raton, FL, 2014), pp. 169–189

98. R. LaComb, O. Nadiarnykh, S. Carey, P.J. Campagnola, Quantitative second harmonic generation imaging and modeling of the optical clearing mechanism in striated muscle and tendon. J. Biomed. Opt. **13**(2), 021109 (2008)

99. A.T. Yeh, B. Choi, J.S. Nelson, B.J. Tromberg, Reversible dissociation of collagen in tissues. J. Invest. Dermatol. **121**, 1332–1335 (2003)

100. N.G. Khlebtsov, I.L. Maksimova, V.V. Tuchin, L. Wang, Introduction to light scattering by biological objects, Chapter 1, in *Handbook of Optical Biomedical Diagnosis*, ed. by V. V. Tuchin, vol. PM107, (SPIE Press, Bellingham, WA, 2002), pp. 31–167

101. T. Yasui, Y. Tohno, T. Araki, Characterization of collagen orientation in human dermis by two-dimensional second-harmonic-generation polarimetry. J. Biomed. Opt. **9**(2), 259–264 (2004)

102. S. Plotnikov, V. Juneja, A.B. Isaacson, W.A. Mohler, P.J. Campagnola, Optical clearing for improved contrast in second harmonic generation imaging of skeletal muscle. Biophys. J. **90**, 328–339 (2006)

103. A. Milgroom, E. Ralston, Clearing skeletal muscle with CLARITY for light microscopy imaging. Cell Biol. Int. **40**(4), 478–483 (2016)

104. E. Olson, M.J. Levene, R. Torres, Multiphoton microscopy with clearing for three dimensional histology of kidney biopsies. Biomed. Opt. Express **7**(8), 3089–3096 (2016)

105. L. Richardson, G. Vargas, T. Brown, L. Ochoa, J. Trivedi, M. Kacerovský, M. Lappas, R. Menon, Redefining 3Dimensional placental membrane microarchitecture using multiphoton microscopy and optical clearing. Placenta **53**, 66–75 (2017)

106. L.F. Ochoa, A. Kholodnykh, P. Villarreal, B. Tian, R. Pal, A.N. Freiberg, A.R. Brasier, M. Motamedi, G. Vargas, Imaging of murine whole lung fibrosis by large Scale 3D microscopy aided by tissue optical clearing. Sci. Rep. **8**, 13348 (2018)

107. C. Zhang, W. Feng, Y. Zhao, T. Yu, P. Li, T. Xu, Q. Luo, D. Zhu, A large, switchable optical clearing skull window for cerebrovascular imaging. Theranostics **8**(10), 2696–2708 (2018)

108. D. Jing, S. Zhang, W. Luo, X. Gao, Y. Men, C. Ma, X. Liu, Y. Yi, A. Budge, B.O. Zhou, Z. Zhao, Q. Yuan, J.Q. Feng, L. Gao, W.-P. Ge, H. Zhao, Tissue clearing of both hard and soft tissue organs with the PEGASOS method. Cell Res. **2**(8), 803–818 (2018)

109. E.A. Susaki, H.R. Ueda, Whole-body and whole-organ clearing and imaging techniques with single-cell resolution: toward organism-level systems biology in mammals. Cell Chem. Biol. **23**(1), 137–157 (2016)

110. P.J. Keller, M.B. Ahrens, Visualizing whole-brain activity and development at the single-cell level using light-sheet microscopy. Neuron **85**(3), 462–483 (2015)

111. P.J. Keller, H.-U. Dodt, Light sheet microscopy of living or cleared specimens. Curr. Opin. Neurobiol. **22**(1), 138–143 (2012)

112. P. Osten, T.W. Margrie, Mapping brain circuitry with a light microscope. Nat. Methods **10**, 515–523 (2013)

113. R. Tomer, L. Ye, B. Hsueh, K. Deisseroth, Advanced CLARITY for rapid and high-resolution imaging of intact tissues. Nat. Protoc. **9**, 1682–1697 (2014)

114. J.M. Girkin, M.T. Carvalho, The light-sheet revolution. J. Opt. **20**, 053002 (2018)

115. A.H. Voie, D.H. Burns, F.A. Spelman, Orthogonal-plane fluorescence optical sectioning: three-dimensional imaging of macroscopic biological specimens. J. Microsc. **170**(3), 229–236 (1993)

116. J. Huisken, J. Swoger, F. Del Bene, J. Wittbrodt, E.H.K. Stelzer, Optical sectioning deep inside live embryos by selective plane illumination microscopy. Science **305**(5686), 1007–1009 (2004)

117. The first commercial ultramicroscope. http://lavisionbiotec.com/2010.html. Accessed 21 Mar 2019

118. P.A. Santi, Light sheet fluorescence microscopy: a review. J. Histochem. Cytochem. **59**(2), 129–138 (2011)

119. L. Silvestri, A. Bria, L. Sacconi, G. Iannello, F.S. Pavone, Confocal light sheet microscopy: micron-scale neuroanatomy of the entire mouse brain. Opt. Express **20**(18), 20582–20598 (2012)

120. B. Lloyd-Lewis, F.M. Davis, O.B. Harris, J.R. Hitchcock, F.C. Lourenço, M. Pasche, C.J. Watson, Imaging the mammary gland and mammary tumours in 3D: optical tissue clearing and immunofluorescence methods. Breast Cancer Res. **18**(1), 127 (2016)

121. A. Greenbaum, K.Y. Chan, T. Dobreva, D. Brown, D.H. Balani, R. Boyce, H.M. Kronenberg, H.J. McBride, V. Gradinaru, Bone CLARITY: clearing, imaging, and computational analysis of osteoprogenitors within intact bone marrow. Sci. Transl. Med. **9**(387), eaah6518 (2017)

122. J.N. Singh, T.M. Nowlin, G.J. Seedorf, S.H. Abman, D.P. Shepherd, Quantifying three-dimensional rodent retina vascular development using optical tissue clearing and light-sheet microscopy. J. Biomed. Opt. **22**(7), 076011 (2017)

123. Y. Henning, C. Osadnik, E.P. Malkemper, EyeCi: optical clearing and imaging of immunolabeled mouse eyes using light-sheet microscopy. Exp. Eye Res. **180**, 137–145 (2019)

124. A.P. Di Giovanna, A. Tibo, L. Silvestri, M.C. Müllenbroich, I. Costantini, A.L. Mascaro, L. Sacconi, P. Frasconi, F.S. Pavone, Whole-brain vasculature reconstruction at the single capilary level. Sci. Rep. **8**, 12573 (2018)

125. M.C. Müllenbroich, L. Silvestri, A.P. Di Giovanna, G. Mazzamuto, I. Costantini, L. Sacconi, F.S. Pavone, High-fidelity imaging in brain-wide structural studies using light-sheet microscopy. eNeuro **5**(6), ENEURO.0124-18.2018 (2018)

126. N. Renier, E.L. Adams, C. Kirst, Z. Wu, R. Azevedo, J. Kohl, A.E. Autry, L. Kadiri, K.U. Venkataraju, Y. Zhou, V.X. Wang, C.Y. Tang, O. Olsen, C. Dulac, P. Osten, M. Tessier-Lavigne, Mapping of brain activity by automated volume analysis of immediate early genes. Cell **165**, 1789–1802 (2016)

127. M.D. Rocha, D.N. Düring, P. Bethge, F.F. Voigt, S. Hildebrand, F. Helmchen, A. Pfeifer, R.H.R. Hahnloser, M. Gahr, Tissue clearing and light-sheet microscopy: imaging the unsectioned adult zebra finch brain at cellular resolution. Front. Neuroanat. **13**, 13 (2019)

128. J. Bürgers, I. Pavlova, J.E. Rodriguez-Gatica, C. Henneberger, M. Oeller, J.A. Ruland, J.P. Siebrasse, U. Kubitscheck, M.K. Schwarz, Light-sheet fluorescence expansion microscopy: fast mapping of neural circuits at super resolution. Neurophotonics **6**(1), 015005 (2019)

Chapter 8
Other Applications of Optical Clearing Agents

8.1 Introduction

The use of OCAs in biological materials can be made with different objectives than increasing their transparency and clearing related methods can have application in other areas than clinical practice and associated research. In Sect. 6.4 an ex vivo method to evaluate the diffusion properties of OCAs in biological tissues was described. Although such method is based on collimated transmittance (T_c) and thickness measurements and has some limitations to be applied freely in vivo, similar evaluation could be made from in vivo tissues using optical coherence tomography or diffuse reflectance measurements.

The evaluation of the diffusion properties for OCA's and other products in ex vivo biological materials has a great interest in other fields like cryogenics, food industry, pharmacology, cosmetics, or poison studies. In the following sections, the various areas where the knowledge gathered in the studies of optical clearing (OC) might be important to study the diffusion of products in biological materials will be explored.

8.2 Dermatology, Cosmetics, and Pharmacology

The application of sodium chloride (NaCl), glucose, or glycerol to the eye to treat corneal edemas has been known for a long time. Glycerol drops can be applied directly to the eye's cornea whenever clouding occurs as a result of epithelial edema. It clears the cornea in 20 or 30 s, and it causes less smarting than glucose or NaCl [1]. Glycerol presents no toxicity when instilled into the conjunctival sac, and the addition of sodium carboxymethylcellulose increases the dehydration effect by increasing its viscosity without decreasing its transparency [2]. Topical glycerol is painful when applied to the eye. For this reason, it is advisable to use it after the

L. M. C. Oliveira, V. V. Tuchin, *The Optical Clearing Method*,
SpringerBriefs in Physics, https://doi.org/10.1007/978-3-030-33055-2_8

application of topical anesthesia. Due to its painful nature and short-time activity, glycerol is not adequate for chronic therapy, but it is useful for epithelial edema clearing to allow visualization with gonioscopy or ophthalmoscopy [3]. For more advanced symptomatic chronic corneal edema, aqueous solution containing 5% of NaCl can be used. Such solute penetrates the epithelium poorly and can therefore attract more easily the diffusible water from the epithelial bullae [4]. A mixed ex vivo and in vivo study was performed with rabbit and human eyes to clear the cornea. For the ex vivo samples, the entire eye tissues have cleared after soaking in nonionic radiological contrast media, such as Amipaque or Omnipaque. For the in vivo study, the cornea became clear, allowing one to see the retina with an ophthalmoscope [5]. No side effects were detected in this study.

Many skin drugs, oils, creams, sun creams, and lotions are topically applied to the human skin with various purposes. The skin is the largest organ in the human body, and its major function is to protect the internal organs from outside bacteria, pollution, or hazardous events. The skin structure is complex, containing superimposed layers. The outermost skin layer is the stratum corneum (SC), which is mostly composed of dead cells. Below the SC we find the epidermis and then the dermis layers [6]. Figure 8.1 shows a schematic of the skin structure.

The outermost skin layer, the SC, is the most important barrier of the skin. It consists of a lipid-protein biphasic structure that has tight cell packing and presents cell membrane keratinization. Its thickness varies between 10 and 20 μm in most areas of the body [6]. Such physiology turns the SC a strong outside barrier to invading organisms or to prevent the penetration of foreign molecules [6–13]. Due to the excellent barrier that SC provides, the delivery of immersion agents to the deeper

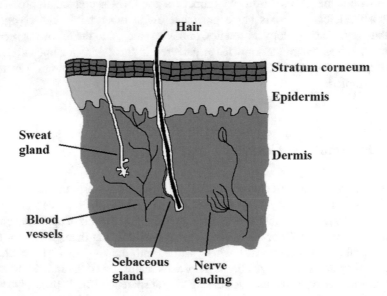

Fig. 8.1 Schematic of skin structure showing the three skin layers (adapted from various images available in the Internet)

layers and water loss by the skin becomes difficult [6, 12]. Dermis and epidermis are permeable skin layers [12], turning the fluxes of water or agents much easier.

Topical and transdermal drug delivery to the skin has been made since ancient times [13]. It is reported in the *Ebers Papyrus*, dated to 1550 BC, that several mixtures and formulations were available for various skin conditions. Other historical references indicate preparations containing hallucinogenic substances that were made to be topically delivered in the inner part of the legs, where permeability is high [13].

Mathematical modeling of transdermal absorption and topical delivery of drugs have been tried since 1940. Several studies revealed that partitioning and solubility of the drugs were very important factors that determine skin penetration [12]. It has been reported [12] that most drug absorption is transcellular in the skin, meaning that noticeable drug absorption between the cells or through sweat pores and hair follicles is unlikely to occur. It was also observed that drug diffusion in the skin is a passive process, and its magnitude depends on the drug itself and on the integrity of the epidermal barrier. Highest penetration in the skin has been observed for drugs with low molecular weight (below 800 Da) and with high water and lipid solubility [12]. Other important aspects on drug penetration are the vehicle that contains the drug and degree of hydration of the SC [12].

By 1960, it was observed that the transport phenomena in the skin can be described by Fick's first law (Eq. 2.11), and a mathematical model describing percutaneous absorption as a passive diffusion process in the vehicle and the membrane layers in series was developed [6]. A paper published in 1961 by Takeru Higuchi [14], presents an elegant set of equations that describe the release rate of drugs from an ointment to the skin. One of these equations relates the amount and release rate of the drug with the diffusion coefficient in the ointment:

$$Q = (2A - C_s)\sqrt{\frac{Dt}{1 + \frac{2(A-C_s)}{C_s}}}. \tag{8.1}$$

In Eq. (8.1), Q is the amount of the drug that was released until the time t per unit area. A represents the concentration of the drug expressed in cm^{-3}, C_s is the solubility of the drug in the external ointment, also in cm^{-3}, and D is the diffusion coefficient of the drug in the ointment [14].

More recently, studies have been made to evaluate the diffusion coefficient of drugs in the skin. If we consider a skin membrane that is exposed to a solute on one side, the fundamental equation that describes steady-state transport through a slab-form skin sample can be derived from Fick's first law of diffusion [13]:

$$Q = \frac{DAT\Delta C_s}{h}, \tag{8.2}$$

where Q is the amount of solute crossing the skin sample of area A, over a time period T. ΔC_s represents the constant concentration gradient across the two surfaces of the skin, D is the diffusion coefficient in the skin sample, and h represents the path length.

Equation (8.2) was derived, assuming that the skin barrier, SC, behaves like a pseudo-homogeneous membrane and that its barrier properties do not change with time or with position [13]. It should be emphasized that the steady-state diffusion is not immediate. The time necessary to reach steady-state diffusion across a homogeneous membrane is given by $h^2/6D$. Equation (8.2) is usually represented in terms of steady-state skin flux, J_{ss}, which is defined as [13]:

$$J_{ss} = \frac{Q}{AT} = \frac{D\Delta C_s}{h}. \qquad (8.3)$$

Various other models of drug diffusion in the skin are presented in literature [13], where additional parameters are used, such as the permeability coefficient, K_p [15], or describing the role of penetration enhancers, like glycerol or urea [16, 17]. Other studies on assisted drug penetration methods like microencapsulation [18] or microneedles [19] have also been reported.

The study of drug diffusion in the skin can be made by studying only D, since it can be related to K_p in the following way [20]:

$$D = \frac{K_p x}{K_m}, \qquad (8.4)$$

where K_m is the membrane partition coefficient, and x is the axis-coordinate in the diffusion direction.

An ex vivo method commonly used to estimate both the partition and the diffusion coefficients is made using the Franz diffusion cell [20–22]. The structure of the Franz cell is represented in Fig. 8.2.

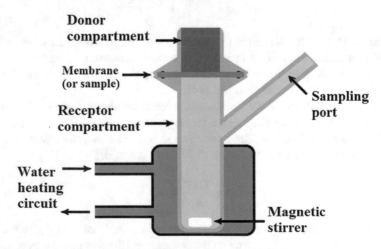

Fig. 8.2 Schematic of a Franz diffusion cell (adapted from Ref. [23])

The Franz diffusion cell contains a donor compartment where the compound for which we want to study diffusion is placed (on top of Fig. 8.2). Due to the gravity force, the compound crosses the tissue sample (or membrane) where we want to determine the diffusion properties. The recipient compartment, that receives the compound under study after crossing the sample, is filled with a medium, which in general is phosphate buffered saline (PBS). Regular time analysis of the receptor medium (through the sampling port) provides time dependence data on the compound concentration increase in the receptor medium. Using Fick's law, the diffusion properties in the sample can be obtained for the tissue sample. A water heating circuit and magnetic stirrer are used on the receptor compartment to keep the solution in the receptor medium homogeneous and to avoid temperature variations.

This method has been widely used to evaluate the diffusion properties of drugs, cosmetics, and creams in ex vivo skin samples [20–22], but it takes some hours to make such evaluation [21].

Most skin creams, ointments, oils, or topical drugs have osmotic properties, meaning that aqueous solutions with different concentrations of these products can be prepared. By using those solutions to treat the skin, thickness and T_c measurements can be acquired during treatments to make the calculations as described in Sect. 6.4. From T_c kinetics we obtain a mean diffusion time, τ, for each treatment and representing it as a function of the product concentration in the solution, we can interpolate to extract the real τ value of the product in the skin sample. Considering a treatment with a product concentration that corresponds to the estimated τ value, by retrieving the sample thickness at that time from thickness kinetics measurements, the calculation of the corresponding diffusion coefficient can be made (see Eq. 6.13).

If a particular skin layer is to be used in this evaluation, the process can be made in a faster way. As referred above, the skin is a multilayered tissue, and different bound/mobile water contents have been evaluated in the various layers [24].

Considering the mobile water for a particular skin layer, we need to acquire T_c and thickness measurements only from a treatment with a solution containing the same water as the mobile water in that skin layer. This way, due to the water balance between the skin sample and the solution, a unique diffusion of the product into the sample occurs, leading immediately to the determination of the τ value of the product in the skin layer.

This method is a robust and fast way to evaluate the diffusion properties of various products in the skin, provided that they have osmotic properties and are soluble in water.

8.3 Tissue Poisoning

Accidental tissue poisoning through topical administration is a great concern, since poisons and other toxic substances are known to be able to diffuse to the inside of the tissues and can be absorbed by tissue cells. The evaluation of the diffusion properties of poisons and toxic compounds in biological materials is thought of major importance, so that treatment procedures can be made as fast and effective as possible.

Toxic chemical agents are commonly used in industry, in agriculture, or in cleaning activities, and accidental contact may occur. Other kinds of hazardous exposure to toxic substances, like skin exposure to poisonous plants or poisonous animals, are also possible.

The human tissues that have major risk of exposure to poisons or toxic agents, like the skin or eye's cornea and sclera, are in contact with the outside of the body. Other inner tissues are also important, since inhalation or ingestion of poisons might also occur, namely, in children. In this case, the evaluation of poison diffusion in tissues from the respiratory or gastric systems is also of interest.

As referred in the previous section, the skin is the largest tissue in the human body, having a wide superficial area. The risk of accidental skin exposure to poison or toxic agents is though high. A recent study was made to evaluate the dermal absorption and skin damage after hydrofluoric acid (HF) exposure [25]. This study reported that HF rapidly penetrates the skin, even in small concentrations, and creates an intradermal reservoir. For HF concentrations above 30%, skin damage progresses into deeper skin layers, as observed in histology photographs. Although the static Franz diffusion cell was used in this study to evaluate HF absorption in the skin, no estimation of the diffusion coefficient for HF in the skin was made.

When the skin is exposed to low volatile organophosphorus compounds, such as pesticides or chemical warfare nerve agents like the VX agent, these may penetrate the skin leading to an uptake by blood circulation. These agents are toxic, and decontamination procedures are performed in those cases to reduce toxic effects. Studies on decontamination of the VX agent have been performed, and good results were obtained for decontamination procedures applied within the first 2 h of the skin exposure to the agent [26]. Time delay after skin exposure is a key factor in these cases. A complementary study performed by the same group [27] reported that if decontamination is initiated within the first 30 min after skin exposure, minimal penetration occurs, and decontamination is more effective. No diffusion properties are reported for these agents, but such evaluation is of significant importance, so that optimized decontamination procedures can be made.

Another toxic compound, similar to VH nerve gas, is the 2-chloroethyl ethyl sulfide (CEES), an alternate and very similar to mustard gas that is responsible for alkylation of proteins. A recent study of CEES in skin samples was made using Franz cells to evaluate and compare the decontamination potential of *dermal decontamination gel* (DDGel) and *reactive skin decontamination lotion* (RSDL) [28]. This study revealed that DDGel is more efficient in CEES removal than RSDL. The evaluation of the diffusion coefficient for CEES in the skin could bring new insight into the reported results and could optimize decontamination doses and procedures.

Methyl chloride is an industrial toxic agent that can be accidentally inhaled or absorbed through the skin. This agent is produced by the reaction of methanol and hydrogen or via the chlorination of methane, and it is used industrially in the production of silicones, butyl rubber, methylene chloride, plastics, pesticides, pharmaceuticals, dyes, resins, polystyrene, and polyurethane foams [29, 30]. It has been reported that inhalation of high concentrations of methyl chloride leads to kidney and liver damage and central nervous system depression, and it may ultimately cause death [31]. Although it is more common to be inhaled and exposure occurs through the lungs and upper airway tissues, methyl chloride can also be absorbed through the skin in amounts that substantially contribute to systemic intoxication [29]. A study has reported that methyl chloride permeates through the skin epidermis when exposure occurs at high atmospheric concentrations within relatively short timeframes [29]. This study revealed that methyl chloride penetration in the skin increases with agent concentration and also with time of exposure. No data is available for the diffusion coefficient of methyl chloride in the skin, but the correspondent air-skin partition coefficient (see Eq. 8.4) has been reported as 13.6 ± 0.5 (dimensionless) [32].

With the recent market boom for electronic cigarettes (e-cig), there is a recent study that makes an alert about nicotine absorption in the skin [33]. This study has reported that nicotine can be absorbed through the skin when the skin is contaminated by e-cig refill liquids. Considering skin samples contaminated with two different refill liquids, skin absorption of nicotine presents no lag time. The observed absorption flux was $0.29 \pm 0.20 \ \mu g/cm^2/h$ for skin samples contaminated with both refill liquids, reducing to $0.19 \pm 0.07 \ \mu g/cm^2/h$, after 10 min washing of the samples. It is important to remember that such absorption can be cumulative over time, leading to oversaturation doses at later time and special caution should be taken regarding children, where tolerated doses are smaller than in adults [33]. The evaluation of the diffusion coefficient of tobacco products in the skin is also of interest to investigate, since no studies have been reported.

There are several other poisonous or toxic products that have been reported to be able to diffuse or to be absorbed in the skin. Some examples of accidental exposure have been reported for hydrogen sulfide and phosphine, which are used in chemical production and agriculture [34]; aconitum alkaloids from aconite (poisonous) plants [35]; copper sulfate, which is used as a fungicide, in photography and in metals industry [36]; or chlorfenapyr, an insecticide commonly used in agriculture for insect and mite control [37]. No data is available for the diffusion properties of these agents in the skin or in other tissues. Such evaluation would provide valuable information to optimize treating or decontamination procedures.

Propylene glycol (PG) is commonly used in optical immersion clearing (OC) studies, sometimes as an OCA, but mostly as an optical clearing enhancer [38, 39]. PG is also used as a solvent for oral, intravenous, or topical pharmaceutical agents. When used in small doses, it is considered safe, but for prolonged periods or in high doses, it may cause toxicity in the central nervous system, cardiac arrhythmia, or other side effects [40]. Using an approximation to the method described in Sect. 6.4, Genin et al. have estimated the diffusion coefficient of PG in rat skin as $(1.35 \pm 0.95) \times 10^{-7} \ cm^2/s$ [41]. Although rat skin was used in this study, this result should not differ much from the corresponding value in human skin.

As e-cig liquid vapor (a mixture of glycerol and PG in different proportions from "soft" with a higher glycerol concentration to "strong" vapor with a higher concentration of PG) recently was suggested as an effective OCA working in a wide spectral range from 200 to 800 nm for the mucous membrane with periodontal lesions, the toxicity of such liquids should be reinvestigated in order to proof protocols for safe biomedical application and evaluate smoking restrictions on e-cigarettes [42].

The evaluation of the diffusion properties of poison and toxic compounds is important, but the same evaluation of chemicals or drugs used in decontamination or treatment procedures is not less important, since optimization of those procedures and drug dosage is also of interest. Considering such treatment or therapeutic procedures, there is a particular case of interest. The use of submicron spheres that encapsulate drugs for follicular diseases or to prevent hair growth in the skin was patented in 2016 [43]. Such spheres are the transport vehicles for the active drugs, and they are constructed in a way that their matrix dissolves after a time delay that corresponds to the travel time to reach the desired skin depth. The evaluation of the diffusion coefficient for these spheres in skin tissue would allow the optimization of drug dosage and delivery [44–46].

Ethylene glycol (EG) is referred as a toxic alcohol, since it originates several pathological changes in various biological tissues [47]. A recent study performed by the group of Luís Oliveira has determined the diffusion properties of EG in skeletal muscle, which may be used for the establishment of treatment protocols to avoid or revert the disruption of the protein chains in the muscle [48].

Considering the respiratory and digestive tract, there are several poisons that can be inhaled or ingested. The major exposure to inhaled gaseous poisons or toxins like carbon monoxide or cyanogen occurs in the lungs, leading to the transfer to the bloodstream that will deliver them to the cells through the entire body. On the other hand, there are certain poisons that can be ingested, and their passage to the bloodstream is made in the liver. Dosage of the poison is again the key factor for the severeness of the effects that will occur. In many cases, fatal doses happen, and the poisoning turns irreversible. There are several cases that have been reported for poisoning through inhalation or ingestion.

One example of poisoning through inhalation that resulted in fatality was reported for a 27-year-old man working on a metal processing factory [49]. This man was clearing an area of the factory, and due to wrong protection equipment, his death resulted from cyanide poisoning. This man was autopsied 3 days after he was found dead, and cyanide contents were measured in different parts of his body. A total of 0.05 µg/mL were found in stomach contents, 7.7 µg/mL in lung tissues, 6.3 µg/mL in heart blood, and 31 µg/mL in the femoral vein blood [49]. These contents show the route of cyanide inside the body—upper respiratory tract—lungs—blood stream—heart—rest of the body.

Another poisoning example occurs through the digestive system. Several household products contain ethanol, caustic soda and chlorine (in bleach), or other toxic products. When these products are not safely guarded, children lead by curiosity may drink these products. The ingestion of those products may induce intoxication and hypoglycemia, and in some cases, it may be fatal. Some comparative statistics have been reported for the ingestion of household products containing ethanol between

2000 and 2010: an increase from 2903 to 18,446 cases [50]. Ethanol is an alcohol used in many applications, including OC, but only in small doses.

Toothpaste also contains some potential toxic agents, such as triclosan, sodium lauryl sulfate (SLS) and PG [51], glycerol [52, 53], diethylene glycol [54, 55], paraformaldehyde [56], or sorbitol [55, 57]. Once again, the dose of such agents in toothpaste and their diffusion ability in tissues are key factors for potential hazardous effects. A study was performed to evaluate the toxicity effects in human gingiva and oral mucosa as a result of exposure to toothpastes containing SLS [58]. This study demonstrated that oral mucosa samples exposed to toothpaste containing SLS developed severe necrosis. Another study on the evaluation of cytotoxicity of toothpaste ingredients on gingival fibroblasts and epithelial cells showed that pastes containing SLS and amine fluoride are more toxic than the ones that contain cocamidopropyl betaine and Steareth-20 [59]. Other types of allergic effects have been reported for a 25-year-old male who systematically used an SLS-containing toothpaste [60]. This patient initially complained of dentinal hypersensitivity and gingival bleeding, and after a series of tests, it was possible to associate other previous symptoms, such as nausea, abdominal gases, bloating, and persistent diarrhea, to the continuous use of toothpastes containing SLS. To evaluate the patient's allergy to SLS, a skin prick test was made. After intradermal injection of 10^{-3} dilution of 1 mg/mL solution of SLS in patient's forearm, an edema of 6 mm in diameter and an erythema of 30 mm in diameter were formed [60]. The detection of diethylene glycol in toothpaste has been reported in a recent study, where about 40% of the pastes analyzed contained not more than 0.1% of this toxic glycol [54]. The method used in this study also detected the presence of other polyethylene glycols in the pastes analyzed. Many counterfeit toothpastes have diethylene glycol as a cheaper replacement for glycerol, and analysis to detect it is performed regularly, due to the large number of deaths occurred before 1937 as a cause of its poisoning [55].

Poisoning through the eye tissues may also occur, since they are also in contact to the environment. Some cases have been reported regarding the contact of toxins with eye tissues. Figure 8.3 shows a schematic of the eye structure.

Considering Fig. 8.3, the eye tissues in contact with the environment are the cornea and the sclera, since the other eye tissues are protected from the outside.

Fig. 8.3 Eye structure (adapted from various images available in the Internet)

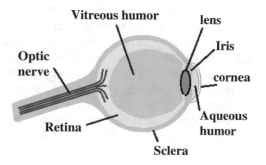

As indicated above, when we presented some cases of skin poisoning, organo-phosphate compounds (the nerve agents) used in warfare are considered as part of the most powerful poisons [61]. Some of these nerve agents are soman, sarin, cyclosarin, tabun, and the already referred VX gas. When someone is exposed to these gases, the irreversible inhibition of acetylcholinesterase occurs, leading to the accumulation of excessive acetylcholine levels in synapses, which causes symptoms like hypersecretions, tremors, status epilepticus, respiratory distress, and eventually death [61]. Both in vapor and in liquid states, the various nerve agents can move into the human body through the skin or through the eyes, but direct ocular contact results in rapid local and systemic toxic effects [61].

Since the eyes are located in the human head, non-intentional eye exposure to toxins or poisons is difficult, unless the poison is in gas state. Many accidental eye exposures occur in industrial locations when adequate protecting goggles are not used [62]. The occupational poisoning cases reported by telephone to the Swedish Poisons Information Centre between 2010 and 2014 are reviewed in Ref. [62]. The eye is at the top of the list with 37% of the reported cases. Poisoning through inhalation comes in second and through the skin comes in third place. The three poison classes that were reported were alkali (excluding ammonia), hydrocarbons, and acids.

Eye exposure to high levels of carbon monoxide or nitrogen dioxide can cause eye irritation [63]. Exposure to tobacco smoke is also dangerous, since it contains substances like ammonia, carbon monoxide, formaldehyde, benzene, nicotine, and various potentially genotoxic or carcinogenic organic compounds [56]. The study of rat's eyes exposure to tobacco smoke concluded that animals that were exposed developed decreased tear stability, ocular surface damage, and altered conjunctival phenotype [64].

The eye tissues may absorb some toxins that were not introduced by direct eye exposure. A case of high arsenic content in the eye of mice after drinking water containing sodium arsenite for several weeks was reported [65]. Having 12 weeks of age, between three and five animals started to drink water containing 0, 10, 50, or 250 ppm of sodium arsenite. For the animals drinking water with 250 ppm of sodium arsenite, the duration of the study was 1 month, and for the others it extended for 6 months. At the end of the studies, the animals were sacrificed and tissues from various organs were collected for mass spectroscopy analysis. The arsenic content in eye tissues was higher than the one observed for the liver, lung, heart, spleen, and brain, and it was similar to the content observed in kidneys.

Many of the poisonous or toxic substances presented in the previous paragraphs have osmotic properties, and their diffusion can be described by Fick's law of free diffusion, similar to the case of OCAs used in the OC treatments. For those particular substances, similar procedure to the one described in Sect. 6.4 can be used to determine the diffusion properties of those substances in the tissues where the exposure occurs. The determination of those properties could bring valuable information to optimize the treatment dosage or decontamination procedures.

8.4 Food Industry

Due to the reversibility of the OC treatments, they can be used to dehydrate food products, with the objective of optimizing preservation.

A major concern in food preservation is dehydration. Conventional food drying, in hot air, is used since ancient times and offers dehydrated products that can have an extended life of a year [66]. Unfortunately, such procedure reduces the quality of the product, and the freeze-drying method is though preferred.

Due to the absence of liquid water and the low temperatures required to freeze the food products, most microbiological reactions and deterioration stop, giving a final product of excellent quality. Some comparative studies were made in the 1990s to compare between the air-drying and freeze-drying methods through the evaluation of quality and financial aspects. Although the freeze-drying is more expensive, the improved quality and life-extent of the preserved foods turned this method preferable [66].

More recently, the osmotic-induced dehydration of food products has gained an increasing interest, since more effective dehydration, extension of product life, and improvement of the nutritional and organoleptic properties of food are obtained [67–70].

Some advantages can be found on osmotic dehydration of food products: (1) quality improvement for color, flavor, and texture; (2) energy efficiency; (3) decrease of packaging size and distribution costs; (4) needless for chemical treatment; and (5) retention of nutrients and product stability during storage [71]. Some studies have evaluated these advantages. In one study, cashew apples treated with an osmotic solution of NaCl and sucrose, after air-drying, revealed better color, firmer texture, astringency, and overall acceptability [72]. Another study demonstrated low shrinkage, better rehydration, and soft raisin-like texture from previously air-dried blueberries after treatment with an osmotic solution of sucrose [73]. Peppers that were dried with whey and sorbitol developed agreeable flavor, color, and texture [74]. Peas were dehydrated in sucrose and citrate solution and developed pleasant flavor, color, and texture without preservatives [75]. Apples and strawberries that were first treated with sucrose solutions, followed by microwave-vacuum dehydration, developed better quality in terms of color, taste, vitamin C, and structure retention [76].

Studies conducted with scanning electron microscopy on the dehydrated apples and strawberries revealed that the cellular structure was better preserved for fruits that were treated with osmotic sucrose [76]. Mechanical response of strawberry was similarly preserved with air-drying osmotic and with freezing-thawing treatments [71]. Kiwi fruit treated with glucose, glycerol, or sucrose solutions presented lower failure forces than fresh fruit, while treatments with calcium lactate increased the failure due to enhanced cell packing and increased cell wall integrity [71].

A traditional preserving method for meat and fish is food salting. It has been reported that this process produces an increase in salt content and a decrease in water content on meat or fish products [77]. When salt grains are applied to meat or fish surfaces in large quantity, they stimulate dehydration and diffuse also into the interstitial space of meat or fish. This is evidence of the two OC mechanisms—tissue dehydration and refractive index (RI) matching. The occurrence of the two OC mechanisms in food products shows that their characterization in terms of determining the diffusion properties for water and agent is of high interest, as in the case of clinical applications, so that optimization of agent dosage and treatment procedures can be performed.

Some studies have been made to evaluate those diffusion properties in food products. One of those studies was performed on apples to evaluate the diffusion coefficients during treatments with aqueous solutions containing sucrose (50% w/w) and salt (10% w/w) for 2, 4, and 8 h at 27 °C [78]. The diffusion coefficients for water, sucrose, and salt as a function of their concentration inside the apples were obtained using a numerical finite differences method. The equation that describes those diffusion coefficients (D) as a function of concentration (ρ) is given as [78]:

$$D_{im}(\rho_i) = a\rho_i^{2} + b\rho_i + c \qquad (i = A, B, C), \qquad (8.5)$$

where m represents the apple medium, A is water, B is sucrose, and C is salt. The estimated values for the constants a, b, and c are presented in Table 8.1.

A similar evaluation study was performed with oyster mushrooms under treatment with sugar-salt solutions with sugar concentrations corresponding to a Brix degree between 45 and 65 and temperatures from 40 to 60 °C [79]. Using Fick's second law of diffusion, the effective water diffusivity was estimated to be within the range $D_w = (3.00–1.59) \times 10^{-5}$ cm^2/s. The estimated diffusivity for the solute was within $D_{sol} = (9.44–10.2) \times 10^{-6}$ cm^2/s [79].

The evaluation of the diffusion coefficients in melon cubes during treatment with aqueous solutions containing sucrose or maltose was made at constant temperature. Melon samples treated with maltose solutions presented higher water loss and lower sugar uptakes than the samples treated with sucrose solutions. Using Fick's law the estimated effective diffusion coefficients for water ranged between 3.93×10^{-5} and 6.45×10^{-5} cm^2/s, while for sucrose they ranged from 7.57×10^{-6} to 3.14×10^{-5} cm^2/s [80]. The effective diffusion coefficients for water and salt (NaCl) in cherry tomato were also estimated using Fick's second law of diffusion, and the estimated values range from 0.43×10^{-5} to 1.77×10^{-5} cm^2/s for water and from 0.04×10^{-5} to 0.54×10^{-5} cm^2/s for salt [81]. Treatments applied to jenipapo fruit samples with sucrose solutions showed that the mass transfer rate for water and solutes and the apparent diffusion coefficients for sucrose depend on sucrose concentration in the osmotic treating solution, while the immersion time does not affect

Table 8.1 Constants for D_{Am}, D_{Bm}, and D_{Cm} coefficients determined with Eq. (8.5) [78]

Coefficients	Constants		
	$a \times 10^{32}$ (cm^8/(kg^2 s))	$b \times 10^{23}$ (cm^5/(kg s))	$c \times 10^{14}$ (cm^2/s)
D_{Am}	−5.27	8.18	−2.88
D_{Bm}	−1.62	3.28	−1.36
D_{Cm}	−1.31	2.61	−1.00

these properties [82]. As we have demonstrated [83], when an osmotic treatment is applied to a biological sample, the estimated diffusion properties correspond to a net flux between the sample and the treating solution. In the majority of cases, such net flux corresponds to a combination of water flux out and agent flux into the sample. According to the method described in Sect. 6.4, means to obtain the individual diffusion properties that correspond to the water flux out and agent flux in are available for ex vivo samples.

It is important to stress that the diffusion coefficient, as well as the diffusion time of water or of an agent in food products, depends not only on the agent concentration in the treating solution but also on temperature. It has been reported that fruits and vegetables treated with glucose or fructose solutions presented increasing diffusivity with increasing temperature [82]. Another study has evaluated the temperature dependence of water loss and sugar gain for papaya cubes and concluded that a maximum water loss and optimum sugar gain are obtained for 37 °C [84]. A similar study reported that optimal dehydration of Amla fruit and sugar uptake is obtained at 30 °C [85]. An optimal temperature of 38 °C was found for the dehydration of mango fruit with sucrose solutions, and an increase in pressure reduces the diffusion coefficient of sucrose [69]. Also in carrots, the estimated diffusion coefficients of water and sucrose have been reported to increase with sucrose concentration in the treating solution and with temperature [86].

The evaluation of water loss and sugar gain in chicken breast meat was evaluated at different temperatures during treatments with aqueous solutions containing NaCl and sucrose [86]. This study shows that water loss increases with increasing treatment duration, with increasing temperature and also with increasing concentration of sucrose in the treating solution. Sugar gain shows similar behavior, but it tends to stabilize with treatment duration. The optimal dehydration conditions were obtained for a treatment with 60% concentration of sugar in the solution, 5 h treatment duration at a temperature of 44 °C. For this case, the water loss was ~48%, and the sugar gain was ~16% [87].

Eugenol is the main component in the essential oil of *Jamaican pepper* plant. It is commonly used in meat industry, as ingredient of cosmetic products and as a therapeutic agent for various diseases [88]. The effective diffusion coefficient values of eugenol in preservative starch-based films placed over mutton meat slices were reported, 1.07×10^{-10} cm^2/s at 10 °C and 1.19×10^{-10} cm^2/s at 15 °C, while the following values were obtained in an edible film containing aloe vera and gelatin, 7.05×10^{-8} to 5.36×10^{-7} cm^2/s at 5 °C and 3.80×10^{-7} to 5.26×10^{-6} cm^2/s at 25 °C [88].

Other osmotic agents, such as NaCl, glucose, fructose, dextrose, maltose, polysaccharide, maltodextrin, corn starch syrup, whey, sorbitol, ascorbic acid, citric acid, calcium chloride, or their combinations have been reported to be suitable for application in food industry. They are used to induce osmotic dehydration of meat and fish products, which gives high-quality products with extended shelf life [89]. Ternary solutions are also used to dehydrate food products. One particular study was made to evaluate the diffusion coefficients of maltodextrin and NaCl in meat during dehydration treatment. Using Fick's second law, the following values were obtained during treatments at 5, 15, and 25 °C: 1.15×10^{-5} to 2.67×10^{-5} cm²/s for maltodextrin and 1.24×10^{-5} to 2.10×10^{-5} cm²/s for NaCl [90]. Similar data has been reported for the diffusion coefficients of water, NaCl, and sucrose in vegetables and fruits. For elephant foot yam (*Amorphophallus* spp.—an African vegetable) cubes under dehydration treatment with NaCl-water solutions, the estimated diffusion coefficients were 7.38×10^{-5} to 8.56×10^{-5} cm²/s for water and 6.78×10^{-5} to 7.92×10^{-5} cm²/s for NaCl. These experiments were conducted with solutions having salt concentrations between 5% and 15% and temperature ranging from 40 to 60 °C. For banana fruit samples treated with water-sucrose-NaCl solutions, containing 0–10% salt and 30–60% sucrose (weight percentages), and temperatures ranging between 25 and 55 °C, the estimated coefficients were 5.19×10^{-6} to 6.47×10^{-6} cm²/s for water, 4.27×10^{-6} to 6.01×10^{-6} cm²/s for sucrose, and 4.32×10^{-6} to 5.42×10^{-6} cm²/s for NaCl [91].

The evaluation of the individual diffusion properties of water and agents in food products, such as fruits, meat, or fish, brings valuable information to optimize agent dosage and reduce treatment time in the food industry. Besides the values presented in literature, there are many others to be studied, and the method described in Sect. 6.4 is easy to implement in food industry and a valuable tool for such market.

8.5   Cryogenics

Tissue and organ preservation is another field where agents with OC properties are widely applied. Considering the preservation temperature, two cases should be considered, such as room temperature (~25 °C) and cryogenic temperatures (in the order of −180 °C and below that).

For the case of tissue preservation at room temperature, there are some successful protocols that have been reported. Subsequently to the successfully preserving of cat corneas in a frozen mixture of alcohol-dry ice for 4 months after dehydration in a mixture of 15% glycerol, some researchers have improved the method to perform preservation at room temperature. They dehydrated human corneas in 95% pure glycerol, without freezing, in sealed tubes to maintain vacuum. The tubes were kept at room temperature for long periods, up to 2 years, and after rehydration, the corneas were used in over 50 patients for lamellar grafting, presenting similar results to the ones observed for fresh harvested corneas [92]. A more recent paper reports that glycerol preservation of acellular corneal tissue is a long-term storage method, which can safely preserve tissues at room temperature up to 5 years after corneal harvesting [93]. This time period is also publicized by a commercial product that is available for this purpose, which allows for corneal storage for transplants [94].

In some particular cases, such as short-time periods, room temperature or refrigerated immersion of skin tissues in glycerol has been proposed as an alternative to preservation in liquid nitrogen [95]. Evaluation of skin samples submitted to these procedures revealed no significant changes in tissue quality under bright-field microscopy or in collagen birefringence. The solution containing 85% glycerol revealed to be the best concentration to optimize water displacement from the skin and to control microbiological growth [95].

A new preservation method for studying purposes only (no transplantations are possible) that works at room temperature has been developed recently—the Elnady technique [96]. In this method, an entire organ, which has been retrieved from animal cadavers, is first submitted to formalin fixation. The second step consists of dehydrating the organ with acetone and glycerol impregnation, and then the organ is cured with cornstarch powder between 3 days and 1 week. At the end, the specimen can be cleaned for any residual cornstarch and can be kept at room temperature for long time without degrading [96]. According to the author, the use of glycerol in this innovative preserving method is related to four properties of glycerol that favor tissue preservation: it is nontoxic, has plasticizing and hygroscopic properties, and presents high penetration rate.

Considering tissue or organ protection at cryogenic temperatures, cryogenic agents with OC properties are also used. Some chemical products are considered as cryoprotective agents (CPAs), since they can dissolve in water to lower its melting point. Some examples of CPAs are ethylene glycol (EG), propylene glycol (PG), glycerol, or dimethyl sulfoxide (DMSO), and they are usually designated as "antifreeze." If we consider a cell culture that is frozen to reach liquid nitrogen temperatures, the growing ice compacts the cells into a continuously decreasing volume of unfrozen liquid as temperature decreases. By adding a CPA, the continuously decreasing liquid volume that contains the cells at any temperature is larger than it would be without the CPA. By providing a larger volume for the unfrozen cells at any given temperature during the freezing process reduces both mechanical damage and excessive salt concentration in the cells [97]. Typical concentrations of 5–15% of a CPA in water are used to allow the survival of a substantial portion of isolated cells after freezing and thawing from liquid nitrogen temperature. The so-called penetrating CPAs are the most commonly used and due to their small-sized molecules, they have the ability to penetrate cell membranes, which allows them to prevent excessive cell dehydration during the freezing process [97].

As suggested in the previous paragraph, the cryogenic preservation method is made by slowly and continuous lowering of temperature. Since tissues or organs do not have space for ice to grow, the growing ice crystals will damage cells and other biological components inside. For an organ to resume function after freezing, a large number of all types of cells in the organ must survive, which is not the case in the majority of cases [97]. To overcome this problem, a new cryogenic procedure, designated as "vitrification," was proposed in 1984 [98]. The novelty of the vitrification method is that in addition to the fast temperature decrease, a concentrated CPA replaces the water in the organ or tissue and also inside the cells, creating a glassy material that minimizes ice crystal formation during the entire cooling process. With the absence of mechanical injury by ice to tissue cells and by keeping salts and other molecules undisturbed in their natural locations, vitrification provides a good solution for the organ and tissue preserving at low temperatures. The only inconvenient of this method is that some CPAs are toxic at low temperatures and may damage tissue cells [97]. A comparative study on Vero cells viability after freezing-thawing with glycerol and DMSO as CPAs was performed. Two similar Vero cell suspensions, which were prepared from the kidney of African green monkeys, were preserved in liquid nitrogen at -196 °C. After 1 year in liquid nitrogen, the cell suspension that used glycerol as CPA showed 70% viability, while the one that used DMSO showed only 60% viability [98]. These results show that DMSO is more toxic to the cells than glycerol, although only 10% concentrations were used in both solutions. Tissue cryopreservation for later use in drug testing is also of interest. Tissue-engineered liver slices (TELSs), which comprised Hep-G2 (human liver cancer cell line) were immersed in calcium alginate gel before cryopreservation at -80 °C for 1–14 days [99]. The objective of such cryopreservation was to compare the effects of posterior treatment with gefitinib (a drug used in different types of cancer) between control and cryopreserved TELSs. Although cell viability before treatment with gefitinib was a little lower in cryopreserved TELSs than in control samples, after treatment all showed similar metabolism and toxicity for the drug [99].

Much is still needed to understand about the toxicity of a CPA at low temperatures, since that harmful behavior is completely different to the one observed at normal temperatures. As an example, we consider that at normal temperatures, PG is not toxic, and EG may metabolize into a poison. At 0 °C, high concentrations of EG are less toxic to cells than PG [97]. Glycerol, DMSO, and other CPAs also have been reported to present adverse reactions at low temperatures [97, 100]. One method to decrease toxicity of CPAs is to mix them with other agents in cryopreservation solutions. Such solutions, which are used both in the freezing or vitrification methods, contain [97]:

1. The carrier solution, which is composed by salts, pH buffer, non-cryoprotectant osmotic agents, nutritive ingredients, or apoptosis inhibitors.
2. Penetrating CPAs, consisting of small molecules that are able to cross cell membranes.
3. Non-penetrating CPAs, which are optional, consist of large molecules that only permeate interstitial locations and do not penetrate cell membranes.
4. Ice blockers, which are also optional, are capable to prevent directly the ice growth, by binding ice or ice nucleators.

The carrier solution prevents the volume variation of cells during the freezing and thawing procedures and helps to control CPA concentration and toxicity in the overall solution. CPAs prevent cell dehydration and ice growth in the freezing method, while in the vitrification method, they completely prevent ice growth. The role of non-penetrating CPAs is also to prevent ice growth, but in the interstitial locations, outside the cells. The non-penetrating CPAs are usually less toxic than penetrating CPAs at the same concentrations. Two examples of non-penetrating CPAs are polyethylene glycol (PEG) and polyvinylpyrrolidone; low molecular weight polyvinyl alcohol, polyglycerol, and biological antifreeze proteins are examples of ice blockers, which are used only in vitrification solutions [97].

Table 8.2 contains a list of CPAs, distributed by classes, which have been experimentally proven as efficient [101].

Table 8.2 Classes of cryoprotective agents

Alcohols and derivatives	Sugars and sugar alcohols	Polymers	Sulfoxides and amides	Amines
Methanol[a]	Glucose[a]	PEG[b]	DMSO[c]	Proline[b]
Ethanol[d]	Galactose[a]	Polyvinylpyrrolidone[a]	Acetamide[a]	Glutamine[a]
Glycerol[e]	Lactose[d]	Dextrans[b]	Formamide[a]	Betaine[a]
PG[b]	Sucrose[b,d]	Ficoll[b]	Dimethyl acetamide[d]	
EG[b]	Trehalose[b]	Hydroxyethyl starch[b]		
	Rafinose[b]	Serum proteins (complex mix)[b]		
	Mannitol[a,d]	Milk proteins (complex mix)[a,d]		
	Sorbitol[d]	Peptones[d]		

[a]Effective to a limited degree in eukaryotic cells
[b]Moderately effective in eukaryotes, often in combination
[c]Highly effective and widely used across all classes of cells
[d]Effective in prokaryotic cells
[e]Very effective in a defined number of cell types

According to the description presented in the previous paragraphs about the freezing and vitrification methods for cryoprotection of tissues and organs, we see that the evaluation of the diffusion properties of osmotic agents in tissues, organs, and their cells is important and necessary to optimize those methods. Such properties depend on temperature [102], but, in general, the diffusion of CPAs in tissues and organs is provided before the freezing process. By permeating tissues and organs at temperatures near 0 °C, toxicity to cells is minimized, and by doing it in steps of 20 min with increasing concentrations of the CPAs, subject shrinkage is minimized [97]. When tissues and organs are retrieved from cryopreservation, the reverse process must be performed, by unloading the cryoprotecting solutions [97, 103, 104].

According to the efficient CPAs indicated in Table 8.2 and the large number of tissues and organs that the human body contains, there are a large number of studies that can be done to evaluate the diffusion properties of such agents near 0 °C. Some works have already been done in the perspective of cryopreservation, but many more must be done in this field. One example of such studies is the evaluation of the diffusion coefficient of DMSO in cartilage tissue. A study made using porcine cartilage at 22 °C reported a diffusion coefficient of 2.3×10^{-6} cm^2/s [105]. Another study evaluated the diffusion coefficient of PG in human skin, fibroid, and myometrium tissues at room temperature and reported the values: 0.6×10^{-6} cm^2/s, 1.2×10^{-6} cm^2/s, and 1.3×10^{-6} cm^2/s, respectively [106]. The permeability coefficients were also reported for some agents in sperm (human or animal): 2.7×10^{-5} cm/s for glycerol (in human), 4.2×10^{-5} cm/s for glycerol (in equine), 3.3×10^{-5} cm/s for erythritol (in bovine), 2.3×10^{-5} cm/s for ribitol (in bovine), and 1.0×10^{-5} cm/s for sorbitol (in bovine) [107]. The diffusion coefficients of EG and PG have also been reported for human ovarian tissue: 1.85×10^{-5} cm^2/s and 1.0×10^{-5} cm^2/s, respectively [108]. The diffusion coefficients of glucose solutions at different osmolarities in normal and diabetic mouse skin have also been reported at 20 °C to vary between 8.83×10^{-7} and 2.87×10^{-6} cm^2/s [109]. Finally, during CPA treatment of tissues or organs, the diffusion coefficient of water going out of the tissue is also important to evaluate, both in tissue and through the solutions going in. The diffusion coefficient of water in hydroxyethyl starch, a stabilizing agent commonly used in the freeze-drying of biological materials, has been reported to be within 1.0×10^{-7} and 2.8×10^{-7} cm^2/s for temperatures between 254 and 293 K [110]. The diffusion coefficient of water flowing through water-glycerol solution (with 0.13 mol fraction of glycerol in solution) was reported as 0.844×10^{-5} cm^2/s at 300 K, while its self-diffusion coefficient is 2.212×10^{-5} cm^2/s at the same temperature [111].

Finalizing this chapter, we see various areas of application of OCAs and most importantly the possibility of conducting ex vivo and in vivo research studies to evaluate their diffusion properties in various materials that are of great importance for many fields of bioengineering.

Also, the analyzed agents, namely, their property to make the object optically transparent, should allow for noninvasive optical methods to monitor the processes of freezing and thawing of the tissue or organ, as well as track changes in the tissue or organ during storage.

References

1. D.G. Cogan, Clearing of edematous corneas by glycerin. Am. J. Ophthalmol. **26**(5), 551 (1943)
2. K.C. Swan, A dehydrating jelly to clear corneal bedewing. AMA Arch. Ophthalmol. **50**(1), 75–77 (1953)
3. B. Duvall, R. Kershner, *Ophthalmic Medications and Pharmacology*, 2nd edn. (SLACK Inc., Thorofare, NJ, 2006), p. 42. Chapter 5
4. C. Costagliola, V. Romano, E. Forbice, M. Angi, A. Pascotto, T. Boccia, F. Semeraro, Corneal oedema and its medical treatment. Clin. Exp. Optom. **96**, 529–535 (2013)
5. D.M. Maurice, Clearing media for the eye. Br. J. Ophthalmol. **71**, 470–472 (1987)
6. V.V. Tuchin, *Optical Clearing of Tissues and Blood* (SPIE Press, Bellingham, WA, 2006)
7. H. Schaefer, T.E. Redelmeier, *Skin Barrier: Principles of Percutaneous Absorption* (Karger, Basel, 1996)
8. F. Pirot, Y.N. Kalia, A.L. Stinchcomb, G. Keating, A. Bunge, R.H. Guy, Characterization of the permeable barrier of human skin in vivo. Proc. Natl. Acad. Sci. **94**, 1562–1567 (1997)
9. I.H. Blank, J. Moloney, A.G. Emslie, I. Simon, C. Apt, The diffusion of water across the stratum corneum as a function of its water content. J. Invest. Dermatol. **82**, 188–194 (1984)
10. T. von Zglinicki, M. Lindberg, G.H. Roomans, B. Forslind, Water and ion distribution profiles in human skin. Acta Derm. Venerol. **73**, 340–343 (1993)
11. J.M. Bradner, Importance of tight junctions in relation to skin barrier function. Curr. Probl. Dermatol. **49**, 27–37 (2016)
12. N. Rutter, Drug absorption through the skin: a mixed blessing. Arch. Dis. Child. **62**, 220–221 (1987)
13. S. Mitragotri, Y.G. Anissimov, A.L. Bunge, H.F. Frasch, R.H. Guy, J. Hadgraft, G.B. Kasting, M.E. Lane, M.S. Roberts, Mathematical models of skin permeability: an overview. Int. J. Pharm. **418**, 115–129 (2011)

14. T. Higuchi, Rate of release of medications from ointment bases containing drugs in suspension. J. Pharm. Sci. **50**, 874–875 (1961)
15. P. Schlupp, T. Blaschke, K.D. Kramer, H.-D. Höltje, W. Mehnert, M. Schäfer-Korting, Drug release and skin penetration from solid lipid nanoparticles and a base cream: a systematic approach from a comparison of three glucocorticoids. Skin Pharmacol. Physiol. **24**, 199–209 (2011)
16. H.A.E. Benson, Transdermal drug delivery: penetration enhancement techniques. Curr. Drug Deliv. **2**, 23–33 (2005)
17. J. Hadgraft, J.W. Hadgraft, I. Sarkany, The effect of glycerol on the percutaneous absorption of methyl nicotinate. Br. J. Dermatol. **87**(1), 30–36 (1972)
18. N. Carreras, C. Alonso, M. Martí, M. Lis, Mass transport model through the skin by microencapsulation system. J. Microencapsul. **32**(4), 358–363 (2015)
19. A. Davidson, B. Al-Qallaf, D.B. Das, Transdermal drug delivery by coated microneedles: geometry effects on effective skin thickness and drug permeability. Chem. Eng. Res. Design **86**(11), 1196–1206 (2008)
20. L. Bartosova, J. Bajgar, Transdermal drug delivery in vitro using diffusion cells. Curr. Med. Chem. **19**, 4671–4677 (2012)
21. H. Rothe, C. Obringer, J. Manwaring, C. Avci, W. Wargniez, J. Eilstein, N. Hewitt, R. Cubberley, H. Duplan, D. Lange, C. Jacques-Jamin, M. Klaric, A. Schepky, S. Grégoire, Comparison of protocols measuring diffusion and partition coefficients in the stratum corneum. J. Appl. Toxicol. **37**, 806–816 (2017)
22. E.A. Genina, A.N. Bashkatov, A.A. Korobko, E.A. Zubkova, V.V. Tuchin, I. Yaroslavsky, G.B. Altshuler, Optical clearing of human skin: comparative study of permeability and dehydration of intact and photothermally perforated skin. J. Biomed. Opt. **13**(2), 021102 (2008)
23. R. Pjanović, R. Stojanović, M. Šajber, J. Veljković, N. Bošković-Vragolović, S. Pejanović, Diffusion of lidocaine hydrochloride from lipid microparticles. Chem. Ind. Chem. Eng. Quart. **15**(1), 33–35 (2009)
24. A.Y. Sdobnov, M.E. Darvin, J. Schleusener, J. Lademann, V.V. Tuchin, Hydrogen bound water profiles in the skin influenced by optical clearing molecular agents- quantitative analysis using confocal Raman microscopy. J. Biophotonics **12**, e201800283 (2019)
25. K. Dennerlein, F. Kiesewetter, S. Kilo, T. Jäger, T. Göen, G. Korinth, H. Drexler, Dermal absorption and skin damage following hydrofluoric acid. Toxicol. Lett. **248**, 25–33 (2016)
26. L. Thors, S. Lindberg, S. Johansson, M. Koch, L. Hägglund, A. Bucht, RSDL decontamination of human skin contaminated with the nerve agent VX. Toxicol. Lett. **269**, 47–54 (2017)
27. L. Thors, M. Koch, E. Wingenstam, B. Koch, L. Hägglund, A. Bucht, Comparison of skin decontamination efficacy of commercial decontamination products following exposure to VX on human skin. Chem. Biol. Interact. **272**, 82–89 (2017)
28. Y. Cao, X. Hui, H. Zhu, A. Elmahdy, H. Maibach, In vitro human skin permeation and decontamination of 2-chloroethyl ethyl sulfide (CEES) using dermal decontamination gel (DDGEL) and reactive skin decontamination lotion (RSDL). Toxicol. Lett. **291**, 86–91 (2018)
29. S. Gaskin, L. thredgold, L. Heath, D. Pisaniello, M. Logan, C. Baxter, Empirical data in support of a skin notation for methyl chloride. J. Occup. Environ. Hyg. **15**(8), 569–572 (2018)
30. R. van Doorn, P.J.A. Borm, C.M. Leijdekkers, P.T. Henderson, J. Reuvers, T.J. van Bergen, Detection and identification of S-methylcysteine in urine of workers exposed to methyl chloride. Int. Arch. Occup. Environ. Health **46**(2), 99–109 (1980)
31. Agency for Toxic Substances and Disease Registry (ATSDR), *Toxicological Profile for Chloromethane* (U.S. Department of Health and Human Services, Public Health Service, Atlanta, GA, 1998)
32. D.R. Mattie, G.D. Bates Jr., G.W. Jepson, J.W. Fisher, J.N. McDougal, Determination of skin: air partition coefficients for volatile chemicals: experimental method and applications. Fundam. Appl. Toxicol. **22**, 51–57 (1994)

33. G. Maina, C. Gastagnoli, G. Ghione, V. Passini, G. Adami, F.L. Filon, M. Grosera, Skin contamination as pathway for nicotine intoxication in vapers. Toxicol. In Vitro **41**, 102–105 (2017)
34. S. Gaskin, L. Heath, D. Pisaniello, R. Evans, J.W. Edwards, M. Logan, C. Baxter, Hydrogen sulphide and phosphine interactions with human skin in vitro: application to hazardous material incident decision making for skin decontamination. Toxicol. Ind. Health **33**(4), 289–296 (2017)
35. T.Y.K. Chan, Aconite poisoning following the percutaneous absorption of Aconitum alkaloids. Forensic Sci. Int. **223**, 25–27 (2012)
36. K.S. Park, J.H. Kwon, S.H. Park, W. Ha, J. Lee, H.C. An, Y. Kim, Acute copper sulfate poisoning resulting from dermal absorption. Am. J. Ind. Med. **61**, 783–788 (2018)
37. S.-K. Han, S.-R. Yeom, S.H. Lee, S.-C. Park, H.-B. Kim, Y.-M. Cho, S.W. Park, A fatal case of chlorfenapyr poisoning following dermal exposure. Hong Kong J. Emerg. Med., 1–4 (2018)
38. X. Guo, Z. Guo, H. Wei, H. Yang, Y. He, S. Xie, G. Wu, X. Deng, Q. Zhao, L. Li, In vivo comparison of the optical clearing efficacy of optical clearing agents in human skin by quantifying permeability using optical coherence tomography. Photochem. Photobiol. **87**(3), 734–740 (2011)
39. Z. Zhi, Z. Han, Q. Luo, D. Zhu, Improve optical clearing of skin in vitro with propylene glycol as a penetration enhancer. J. Innov. Opt. Health Sci. **2**(3), 269–278 (2009)
40. T.Y. Lim, R.L. Poole, N.M. Pageler, Propylene glycol toxicity in children. J. Pediatr. Pharmacol. Ther. **19**(4), 277–282 (2014)
41. V.D. Genin, A.N. Bashkatov, E.A. Genina, V.V. Tuchin, Measurement of diffusion coefficient of propylene glycol in skin tissue. Proc. SPIE **9448**, 94480E (2015)
42. A.A. Selifonov, V.V. Tuchin, Kinetics of optical properties on selected laser lines of human periodontal gingiva when exposed to glycerol-propylene glycol mixture, in International Symposium FLAMN-19 (Fundamentals of Laser Assisted Micro- & Nanotechnologies), Symposium Program, Paper PS3-C02-9, St. Petersburg, 30 June–4 July, 2019, p.71. https://flamn.ifmo.ru/docs/Program_Symposium_FLAMN_-_19.pdf
43. S.D. Sheffer, H.L.R. Cooper, N. Pologruto, Delivery of pharmaceutical active ingredients through the skin and hair follicles into dermis and transdermal delivery, US Patent No. US2016/0361264 A1, 15 Dec 2016
44. E.A. Genina, Y.I. Svenskaya, I.Y. Yanina, L.E. Dolotov, N.A. Navolokin, A.N. Bashkatov, G.S. Terentyuk, A.B. Bucharskaya, G.N. Maslyakova, D.A. Gorin, V.V. Tuchin, G.B. Sukhorukov, In vivo optical monitoring of transcutaneous delivery of calcium carbonate microcontainers. Biomed. Opt. Express **7**(6), 2082–2087 (2016)
45. I.Y. Yanina, N.A. Navolokin, Y.I. Svenskaya, A.B. Bucharskaya, G.N. Maslyakova, D.A. Gorin, G.B. Sukhorukov, V.V. Tuchin, Morphology alterations of skin and subcutaneous fat at NIR laser irradiation combined with delivery of encapsulated indocyanine green. J. Biomed. Opt. **22**(5), 055008 (2017)
46. Y.I. Svenskaya, E.A. Genina, B.V. Parakhonskiy, E.V. Lengert, E.E. Talnikova, G.S. Terentyuk, S.R. Utz, D.A. Gorin, V.V. Tuchin, G.B. Sukhorukov, A simple non-invasive approach toward efficient transdermal drug delivery based on biodegradable particulate system. ACS Appl. Mater. Interfaces **11**(19), 17270–17282 (2019)
47. S.R. White, Toxic alcohols, in *Rosen's Emergency Medicine: Concepts and Clinical Practice*, ed. by J. A. Marx, R. S. Hockberger, R. M. Walls, vol. 2, 7th edn., (Elsevier, Philadelphia, PA, 2010), pp. 2001–2009
48. L.M. Oliveira, M.I. Carvalho, E.N. Nogueira, V.V. Tuchin, Diffusion characteristics of ethylene glycol in skeletal muscle. J. Biomed. Opt. **20**(5), 051019 (2015)
49. S. Seidl, B. Schwarze, P. Betz, Lethal cyanide inhalation with post-mortem trans-cutaneous cyanide diffusion. Leg. Med. **5**, 238–241 (2003)
50. P. Rayar, S. Ratnaplan, Pediatric ingestions of house hold products containing ethanol: a review. Clin. Pediatr. **52**(3), 203–209 (2012)

51. https://articles.mercola.com/sites/articles/archive/2015/09/09/toxic-toothpaste-ingredients. aspx. Accessed 22 Mar 2019
52. S.S. Konstantinović, B.R. Danilović, J.T. Ćirić, S.B. Ilić, D.S. Savić, V.B. Veljković, Valorization of crude glycerol from biodiesel production. Chem. Ind. Chem. Eng. Q. **22**(4), 461–489 (2016)
53. V.K. Garlapati, U. Shankar, A. Budhiraja, Bioconversion technologies of crude glycerol to value added industrial products. Biotech. Rep. **9**, 9–14 (2016)
54. F. Hernández, M. Ibáñez, J.V. Sancho, Fast determination of toxic diethylene glycol in toothpaste by ultra-performance liquid chromatography – time of flight mass spectrometry. Anal. Bioanal. Chem. **391**, 1021–1027 (2008)
55. S. Barry, J.-C. Wolff, Investigation into the quantitative analysis of diethylene glycol in toothpaste by direct analysis in real time mass spectrometry. Rapid Commun. Mass Spectrom. **30**, 1829–1834 (2016)
56. M. Özgöz, H. Yağiz, Y. Çiçek, A. Tezel, Gingival necrosis following the use of a paraformaldehyde-containing paste: a case report. Int. Endod. J. **37**, 157–161 (2004)
57. G.N. Teke, N.G. Enongene, A.R. Tiagha, In vitro antimicrobial activity of some commercial toothpastes. Int. J. Curr. Microb. Appl. Sci. **6**(1), 433–446 (2017)
58. B.V. Vannet, B. De Wever, E. Adriaens, F. Ramaeckers, P. Bottenberg, The evaluation of sodium lauryl sulphate in toothpaste on toxicity on human gingiva and mucosa: a 3D in vitro model. Dentistry **5**(9), 325-1–325-5 (2015)
59. B. Cvikl, A. Lussi, R. Gruber, The in vitro impact of toothpaste extracts on cell viability. Eur. J. Oral Sci. **123**, 179–185 (2015)
60. M. Ersoy, J. Tanalp, E. Ozel, R. Cengizlier, M. Soyman, The allergy of toothpaste: a case report. Allergol. Immunopathol. **36**(6), 368–370 (2008)
61. T.H. Figueiredo, J.P. Apland, M.F.M. Braga, A.M. Marini, Acute and long-term consequences of exposure to organophosphate nerve agents in humans. Epilepsia **59**(S2), 92–99 (2018)
62. L. Schenk, K. Feychting, A. Annas, M. Öberg, Records from the Swedish poisons centre as a means for surveillance of occupational accidents and incidents with chemicals. Safety Sci. **104**, 269–275 (2018)
63. P.D. Creswell, J.G. Meiman, H. Nehls-Lowe, C. Vogt, R.J. Wozniak, M.A. Werner, H. Anderson, Exposure to elevated carbon monoxide levels at an indoor ice arena – Wisconsin, 2014. Morb. Mortal. Wkly. Rep. **64**(45), 1267–1270 (2015)
64. T. Kojima, M. Dogru, A. Higuchi, T. Nagata, O.M.A. Ibrahim, T. Inaba, K. Tsubota, Protection from acute tobacco smoke exposure evidence from Nrf2 knockout mice. Am. J. Pathol. **185**(3), 776–785 (2015)
65. N.J. Kleiman, A.M. Quinn, K.G. Fields, V. Slavkovich, J.H. Graziano, Arsenite accumulation in the mouse eye. J. Toxicol. Environ. Health A **79**(8), 339–341 (2016)
66. C. Ratti, Hot air and freeze-drying of high-value foods: a review. J. Food Eng. **49**, 311–319 (2001)
67. M.R. Khan, Osmotic dehydration technique for fruits preservation – a review. Pak. J. Food Sci. **22**(2), 71–85 (2012)
68. R.S.F. Filho, R.P. Gusmão, W.P. Silva, J.P. Gomes, E.V.C. Filho, A.A. El-Aouar, Osmotic dehydration of pineapple stems in hypertonic sucrose solutions. Agric. Sci. **6**, 916–924 (2015)
69. A. Ciurzyńska, H. Kowalska, K. Czajkowska, A. Lenart, Osmotic dehydration in production of sustainable and healthy food. Tends Food Sci. Tech. **50**, 186–192 (2016)
70. I. Ahmed, I.M. Qazi, S. Jamal, Developments in osmotic dehydration technique for the preservation of fruits and vegetables. Innov. Food Sci. Emerg. Technol. **34**, 29–43 (2016)
71. M.S. Rahman, Osmotic dehydration of foods. Chapter 19, in *Handbook of Food Preservation*, ed. by M. S. Rahman, 2nd edn., (Taylor & Francis Group LLC, CRC Press, Boca Raton, FL, 2007), pp. 433–446
72. G. Bidaisee, N. Badrie, Osmotic dehydration of cashew apples (Anacardium occidentale L.): quality evaluation of candied cashew apples. Int. J. Food Sci. Technol. **36**, 71–78 (2001)

73. M.H. Kim, R.T. Toledo, Effect of osmotic dehydration and high temperature fluidized bed drying on properties of dehydrated rabbit eye blueberries. J. Food Sci. **52**(4), 980–989 (1987)
74. D. Torreggiani, Technical aspects of osmotic dehydration in foods, in *Food Preservation by Moisture Control. Fundamentals and Applications*, ed. by G. V. Barbosa-Canovas, J. Welti-Chanes, (Technomic Publishing, Lancaster, PA, 1995), pp. 281–304
75. F.K. Ertekin, T. Cakaloz, Osmotic dehydration of peas II. Influence of osmosis on drying behavior and product quality. J. Food Process. Preserv. **20**, 105–119 (1996)
76. U. Erle, H. Schubert, Combined osmotic and microwave-vacuum dehydration of apples and strawberries. J. Food Eng. **49**, 193–199 (2001)
77. A. Chiralt, P. Fito, J.M. Barat, A. Andrés, C. González-Martínez, I. Escriche, M.M. Camacho, Use of vacuum impregnation in food salting process. J. Food Eng. **49**, 141–151 (2001)
78. S.M. Monnerat, T.R.M. Pizzi, M.A. Mauro, F.C. Menegalli, Osmotic dehydration of apples in sugar/salt solutions: concentration profiles and effective diffusion coefficients. J. Food Eng. **100**, 604–612 (2010)
79. H.G. Ramya, S. Kumar, S. Kapoor, Optimization of osmotic dehydration process for oyster mushrooms (Pleurotus sajor-caju) in sodium chloride solution using RSM. J. Appl. Nat. Sci. **6** (1), 152–158 (2014)
80. C.C. Ferrari, M.D. Hubinger, Evaluation of the mechanical properties and diffusion coefficients of osmodehydrated melon cubes. Int. J. Food Sci. Technol. **43**, 2065–2074 (2008)
81. P.M. Azoubel, F.E.X. Murr, Mass transfer kinetics of osmotic dehydration of cherry tomato. J. Food Eng. **61**, 291–295 (2004)
82. A.K. Yadav, S.V. Singh, Osmotic dehydration of fruits and vegetables: a review. J. Food Sci. Technol. **51**(9), 1654–1673 (2014)
83. I. Carneiro, S. Carvalho, R. Henrique, L.M. Oliveira, V.V. Tuchin, A robust ex vivo method to evaluate the diffusion properties of agents in biological tissues. J. Biophotonics **12**, e201800333 (2019). https://doi.org/10.1002/jbio.201800333
84. S.K. Jain, R.C. Verna, L.K. Murdia, H.K. Jain, Optimization of process parameters for osmotic dehydration of papaya cubes. J. Food Sci. Technol. **48**(2), 211–217 (2011)
85. D. Tiroutchevalme, V. Sivakumar, J.P. Maran, Mass transfer kinetics during osmotic dehydration of AMLA (Emblica officinalis L.) cubes in sugar solution. Chem. Ind. Chem. Eng. Q. **21**(4), 547–559 (2015)
86. N.K. Rastigi, K.S.M.S. Raghavarao, Function of temperature and concentration during osmotic dehydration. J. Food Eng. **34**, 429–440 (1997)
87. I. Filipović, B. Ćurčić, V. Filipović, M. Nićetin, J. Filipović, V. Knežević, The effects of technological parameters on chicken meat osmotic dehydration process efficiency. J. Food Process. Preserv. **41**, e13116-1–e13116-7 (2016)
88. N.L. Flores-Martínez, M.C.I. Pérez-Pérez, J.M. Oliveros-Muñoz, M.L. López-González, H. Jiménez-Islas, Estimation of diffusion coefficients of essential oil of Pimenta dioica in edible films formulated with aloe vera and gelatin, using Levenberg-Marquardt method. Rev. Mexicana de Ingeniería Química **17**(2), 485–506 (2018)
89. M. Hadipernata, M. Ogawa, Mass transfer and diffusion coefficient of D-Allulose during osmotic dehydration. J. Appl. Food Technol. **3**(2), 6–10 (2016)
90. D. Dimakopoulou-Papazoglou, E. Katsanidis, Mass transfer kinetics during osmotic processing of beef meat using ternary solutions. Food Bioprod. Process. **100**, 560–569 (2016)
91. Sangeeta, B.S. Hathan, Studies on mass transfer and diffusion coefficients in elephant foot yam (Amorphophallus SPP.) during osmotic dehydration in sodium chloride solution. J. Food Process Preserv. **40**, 521–530 (2016)
92. J.H. King, W.M. Townsend, The prolonged storage of donor corneas by glycerine dehydration. Trans. Am. Ophthalmol. Soc. **82**, 106–110 (1984)
93. N. Gupta, P. Upadhyay, Use of glycerol-preserved corneas for corneal transplants. Ind. J. Ophthalmol. **65**, 569–573 (2017)
94. http://www.globalsightnetwork.org/surgeons/glycerolplus-cornea-products. Accessed 22 Apr 2019

95. M.R. Herson, K. Hamilton, J. White, D. Alexander, S. Poniatowski, A.J. O'Connor, J.A. Werkmeiter, Interaction of preservation methods and radiation sterilization in human skin processing, with particular insight on the impact of the final water content and collagen disruption. Part I: process validation, water activity and collagen changes in tissues cryopreserved or processed using 50, 85 or 98% glycerol solutions. Cell Tissue Bank. **19**, 215–217 (2018)

96. F.A. Elnady, The Elnady technique: an innovative new method for tissue preservation. ALTEX **33**(3), 237–242 (2016)

97. B. Wowk, How cryoprotectants work. Cryonics **28**(3), 3–7 (2007). ed. by J. Chapman, Alcor Life Extension Foundation, Scottsdale, AZ

98. M.S.I. Siddiqui, M. Giasuddin, S.M.Z.H. Chowdhury, M.R. Islam, E.H. Chowdhury, Comparative effectiveness of dimethyl sulphoxide (DMSO) and glycerol as cryoprotective agent in preserving Vero cells. Bangl. Veterin. **32**(2), 35–41 (2015)

99. R. Chen, B. Wang, Y. Liu, R. Lin, J. He, D. Li, A study of cryogenic tissue-engineered liver slices in calcium alginate gel for drug testing. Cryobiology **82**, 1–7 (2018)

100. G.M. Fahy, D.R. MacFarlane, C.A. Angell, H.T. Meryman, Vitrification as an approach to cryopreservation. Cryobiology **21**(4), 407–426 (1984)

101. G.D. Elliot, S. Wang, B.J. Fuller, Cryoprotectants: a review of the actions and applications of cryoprotective solutes that modulate cell recovery from ultra-low temperatures. Cryobiology **76**, 74–91 (2017)

102. P. Kilbride, G.J. Morris, Viscosities encountered during the cryopreservation of dimethyl sulphoxide systems. Cryobiology **76**, 92–97 (2017)

103. X. Zhou, X.M. Liang, J. Wang, P. Du, D. Gao, Theoretical and experimental study of a membrane-based microfluidics for loading and unloading cryoprotective agents. Int. J. Heat Mass Transfer **127**, 637–644 (2018)

104. T.A. Takroni, H. Yu, L. Laouar, A.B. Adesida, J.A.W. Elliott, N.M. Jomha, Ethylene glycol and glycerol loading and unloading in porcine meniscal tissue. Cryobiology **74**, 50–60 (2017)

105. A. Abazari, J.A.W. Elliott, L.E. McGann, R.B. Thompson, MR spectroscopy measurement of the diffusion of dimethyl sulfoxide in articular cartilage and comparison to theoretical predictions. Osteoart. Cartil. **20**, 1004–1010 (2012)

106. J.D. Benson, A.Z. Higgins, K. Desai, A. Eroglu, A toxicity cost function approach to optimal CPA equilibration in tissues. Cryobiology **80**, 144–155 (2018)

107. J.G. Alvarez, B.T. Storey, Evidence that membrane stress contributes more than lipid peroxidation to sublethal cryodamage in cryopreserved human sperm: glycerol and other polyols as sole cryoprotectant. J. Androl. **14**(3), 199–209 (1993)

108. G.D.A. Gastal, B.G. Alves, K.A. Alves, S.O. Paiva, S.G.S. de Tarso, G.M. Ishak, S.T. Bashir, E.L. Gastal, Effects of cryoprotectant agents on equine ovarian biopsy fragments in preparation for cryopreservation. J. Equine Vet. Sci. **53**, 86–93 (2017)

109. D.K. Tuchina, R. Shi, A.N. Bashkatov, E.A. Genina, D. Zhu, Q. Luo, V.V. Tuchin, Ex vivo optical measurements of glucose diffusion kinetics in native and diabetic mouse skin. J. Biophotonics **8**(4), 332–346 (2015)

110. G. Spieles, T. Marx, I. Heschel, G. Rau, Analysis of desorption and diffusion during secondary drying in vacuum freeze-drying of hydroxyethyl starch. Chem. Eng. Process. **34**, 351–357 (1995)

111. L. Weng, S.L. Stott, M. Toner, Exploring dynamics and structure of biomolecules, cryoprotectants, and water using molecular dynamics simulations: implications for biostabilization and biopreservation. Ann. Rev. Biomed. Eng. **21**, 1–31 (2019)

Chapter 9
Future Perspectives of the Optical Clearing Method

9.1 Optical Clearing: A Successful Technique in Tissue Optics

With a long history that goes back to the pioneer discoveries of Spalteholz in 1911 and 1914 [1, 2], the optical immersion clearing (OC) method has gained a great interest in the last 30 years [3]. By the end of the twentieth century, and with the arrival of low-cost instruments, the development or improvement of light-based techniques for clinical diagnosis or therapy has been significant. Considering tissue imaging as an example, and comparing between the traditional and the light-based methods, the latter are simpler and safer to use, do not use ionizing radiation, and are cheaper and in most cases noninvasive. The application of such methods can provide information on tissue physiology and allows for pathology differentiation. In fact, the application of light-based techniques to diagnose or treat cancer and other pathologies has grown considerably in recent years [3–7].

The light-based methods have limited light penetration in tissues due to the high refractive index (RI) mismatch between tissue components, which results in strong light scattering. To overcome such drawback, the application of OC treatments in tissues has been progressively used as a way to increase tissue transparency, providing improved light penetration depth in general and improved contrast or resolution in imaging methods [8, 9].

Using some specialized search engines on the Internet, we looked for the number of publications since 1980 that contain the term "tissue optical clearing" and obtained the data presented in Fig. 9.1.

The blue line in Fig. 9.1 shows the average of publications over time, considering PubMed, Google Scholar, and Web of Science.

Considering the increasing number of publications regarding OC and the state of the art developed so far, new lines of research intend to provide new developments in this technique. We will present some expected developments in the following sections.

L. M. C. Oliveira, V. V. Tuchin, *The Optical Clearing Method*,
SpringerBriefs in Physics, https://doi.org/10.1007/978-3-030-33055-2_9

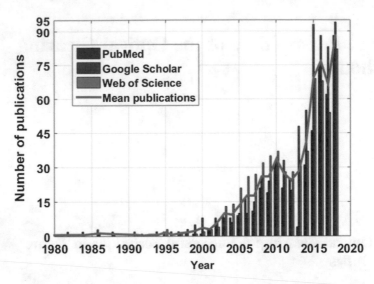

Fig. 9.1 Number of publications about tissue optical clearing from 1980 to 2018

9.2 Future Developments in Clearing Agents and Protocols

The OC potential of chemicals has been studied in many applications as referred in Chap. 3. We also presented in Chap. 3 the desirable characteristics for an agent, in terms of absorption and dispersion, to produce an increase in tissue transparency. In Chap. 7, when presenting the cooperation of OC with imaging methods, we referred several clearing protocols that are adopted for tissue imaging. Both the selection of OCAs and OC protocols are important for any particular situation. To obtain a desirable and reversible transparency in any biological material, we should also keep in mind the OC mechanisms of dehydration, refractive index (RI) matching, and protein dissociation as indicated in Chap. 4.

The discovery of new OCAs may occur, as it was recently demonstrated by Ueda's group [10], where more than 1600 chemicals were screened as potential OCAs by a high-throughput evaluation system for each chemical process. Some widely used products around us unexpectedly may serve as effective OCAs as it was recently proved for electronic cigarette (e-cig) liquid vapor [11] and was suggested to use lactulose as a potential OCA. The e-cig liquid vapor is a mixture of glycerol and propylene glycol, and lactulose is a synthetic disaccharide (4-O-β-D-galactopyranosyl-D-fructofuranose), and they both present increasing RI with decreasing wavelength, as represented in Fig. 3.1. Lactulose is used in solution form for oral or rectal administration for treatment and prevention of portal-systemic encephalopathy [12].

Considering the dehydration and RI matching mechanisms, it is expected that combinations of agents to optimize both mechanisms in a single treatment will be of great interest in the near future. With the objective to clear in vivo tissues, the development of clearing protocols is also highly desirable. The establishment of OCA dosage and application procedure for a variety of tissues and organs in the human body should be considered. For external tissues like the skin, topical application of an OCA may be sufficient, but for internal tissues or organs, such as muscle, gastric, or cardiovascular tissues, other treatment protocols, such as OCA delivery by endoscopy or injection, should be considered and established. However, another prospective strategy was recently proposed for brain-clearing in in vivo microscopy, which is based on glycerol oral administration via drinking glycerol-water solutions, named as MAGICAL (Magical Additive Glycerol Improves Clear Alive Luminescence) [13]. Glycerol is often used for per-oral treatments during stroke, heart attack, edema, and in sport medicine to regulate water balance in the body [14]. As a small molecule, glycerol is transported across the cell membrane by the aquaporin (AQP) family of passive channels [15]. Therefore glycerol digesting may help to provide OC of internal organs due to dehydration and RI matching mechanisms. Evidently, any metabolic OCAs, such as glucose that saturates tissues of diabetic patients, make them more optically transparent. It was found recently that the skin of animals with alloxan-induced diabetes is more optically transparent than of nondiabetic animals [16, 17]. These two examples for glycerol and glucose evidently demonstrate perspectives for in vivo OC of internal organ tissues.

Also, one particular case of interest has been referred in Chap. 7—the clearing of the skull [18–20]. The clearing of cranial bone to image cerebral vasculature or even to monitor brain activity is of high interest, and clearing protocols for the skull should be optimized and established.

9.3   Future Developments in Tissue Spectroscopy

Considering tissue spectroscopy, many studies have been previously performed during optical clearing, but in the majority of these cases, they were made with excised tissue samples. The objective now is to perform similar studies that can produce valuable clinical information, but for the in vivo situation. The most common spectroscopy measuring setup and the most sensitive to variations in tissues during OC treatments is the collimated transmittance (T_c) setup.

The T_c setup is not the best choice to perform measurements from in vivo tissues, especially if we want the procedure to be noninvasive or minimally invasive. The alternative is to select a reflectance measuring setup that can be applied with all the instrumentation located outside the human body and has the same sensitivity as in the T_c setup. The diffuse reflectance (R_d) measurement setup is the perfect choice, since during tissue clearing, scattering reduces, and a corresponding reduction of the R_d signal (see Sect. 5.2.1) can provide valuable clinical information. Considering a specific wavelength band from the R_d spectrum, we can use this setup to estimate the diffusion time for an agent in a tissue and complementary thickness variation data available for that tissue; the diffusion coefficient can also be calculated. The thickness kinetics for a tissue during OC treatment can be obtained from OCT imaging, as referred in Chap. 7. Similarly to what has been presented in Sect. 6.3, the R_d setup can provide an alternative to T_c measurements to calculate the kinetics of the optical properties for in vivo tissues under OC treatments. Such estimation has high interest, since fluid dynamics involved in the dehydration and in the RI matching mechanisms can be used to characterize the treatment and they are different between the in vivo and the ex vivo tissues.

In Sect. 6.5 we presented the creation of two optical windows in colorectal muscle during OC treatments with water-glycerol solutions. Considering such discovery, it is expected that these windows can be used in the near future with an R_d measuring setup to perform diagnosis or treatment procedures. Additionally to reflectance spectroscopy measurements, the use of Raman spectroscopy, fluorescence spectroscopy, or imaging techniques can also benefit with the creation of these two UV windows for diagnostic or treatment purposes.

9.4   Future Developments in Tissue Imaging

According to the recent research examples presented in Chap. 7, much has been done so far with the application of OC treatments to improve the optical imaging methods. Considerable improvements in the imaging probing depth, image contrast, and resolution have been obtained with the application of various clearing protocols. Although we have presented mostly examples of ex vivo tissue imaging using techniques such as OCT and fluorescence microscopy, these techniques as well as speckle contrast imaging can be used to image in vivo tissues. In the examples presented in Chap. 7, to obtain the improved contrast, resolution, and probing depth, it was necessary to submit the tissues to fixation, staining, and OC protocols that took several days. With the objective of acquiring images from in vivo tissues, a significant effort to develop new clearing protocols must be done. Only this way, similar quality of the acquired images can be obtained in the in vivo situation, turning the optical imaging methods a serious alternative to traditional X-ray radiation imaging methods. The possibility of creating 3D images from large in vivo tissue volumes or organs with the application of tissue OC will provide in the future a serious noninvasive alternative to traditional histological methods and will facilitate the establishment of a diagnostic.

Due to the fast image acquisition, the high contrast produced with OC treatments, and the possibility of 3D reconstruction, the light-sheet method is highly promising for the future. If it can be adapted for in vivo image acquisition from internal organs or thick tissues, it will allow one in the future to perform optical biopsy and noninvasive histological analysis. The major problem to adapt the light-sheet imaging method to image in vivo tissues is the necessity of side illumination with one or two lasers (see Fig. 7.5). Such instrumental setup is optimal to image excised tissue volumes at the microscope, but impracticable for the in vivo situation. Research efforts must be performed to adapt this imaging technique to the in vivo situation. The laser beams used to illuminate the tissue volume might need to pass through adjacent tissues, meaning that a working in vivo light-sheet setup and protocol must account for the OC of those adjacent tissues. The capability of OC large tissue volumes or small organs has already been demonstrated [21–23], but it takes a considerable time and is performed in stages, where the OCA concentration in the various solutions increases slowly to avoid significant volume variations. A reduction in the OC time is also necessary for in vivo application, meaning that new OCAs or clearing protocols are necessary.

9.5 Future Developments in Food Industry and Other Areas

According to the several examples presented in Chap. 8, the OC treatments might be useful to other applications beyond the increase of tissue transparency for diagnostic or treatment purposes. Such application of OC treatments in other fields can be done with the objective to evaluate the diffusion properties of agents, poisons, creams, gels, or other products in various biological tissues. With the publication of the method described in Sect. 6.4 and its in vivo analogue with R_d measurements, we expect that OC treatments may be used in other fields to acquire such diffusion properties, which are necessary for various applications.

The study of creams, lotions, or gels' diffusion in the skin is of major interest for pharmacology and for the cosmetic industry. The evaluation of poison diffusion in skin and other biological tissues is also of high importance for the establishment of legislation and development of optimized treatment procedures. In the cryopreservation industry, the evaluation of the diffusion properties of various cryoprotecting agents in various tissues is mandatory to select appropriate agent dosage to avoid cell damage and establishment of optimized thawing procedures. We expect that these industries make use of the methods described in Sect. 6.4 to obtain the diffusion properties of agents and optimize their processes.

In food industry, the OC treatments are already being applied as we have indicated in Sect. 8.4. The application of sugars or sugar solutions to food products provides tissue dehydration with the purpose of optimized long-term preservation and also improves organoleptic properties of the food to be preserved. Such practice is well known today, and the application of OC treatments has already been tried in meat, fruits, and vegetables. There is a large variety of food products with different nature, and the diffusion of sugars, alcohols, or other agents is different in each biological product. This means that research will have to be performed to evaluate the diffusion properties of agents in various food products that require long-term preservation. Additional research to evaluate a particular change in the organoleptic properties of a certain food product is also necessary. As example, some products can change their sweetness or acidity, depending on the amount of sugar, vinegar, or other products used to perform their preparation for preservation or during the thawing process.

The immersion OC approach is opening a unique opportunity for microscope-enabled plant research with a high accuracy [24–26]. There were described a nondestructive OC technique that is compatible with immunocytochemistry and common fluorescent probes, including colored proteins such as GFP, enhances transmission of light through plant samples, and preserves their fluorescence [24]. As an OCA a composition of 6 M urea, 30%-glycerol (v/v), and 0.1% Triton X-100 (v/v) dissolved in sterile water was used and as a permeability enhancer—75%-ethanol. The OCA was applied at room temperature during 2 days or up to a few weeks to make plant structure enough transparent in dependence of plant species under investigation.

As many kinds of plants, natural wood is completely nontransparent because of strong RI mismatch between the cellulose (~1.53) and air (~1.00). Wood tissue is structured in micron-sized channels that scatter light mostly caused by their filling by air and absorb light in the visible range caused by high lignin concentration (20–30 wt%) [27, 28]. To suppress absorption and scattering of wood samples, different delignification technologies and RI matching polymer fillings are in use [27, 28]. One of the wood OC technologies is presented in Ref. [28], where a key point is the use of H_2O_2 steam for delignification. The delignification procedure works as an enhancer of OCA permeability because of providing better tissue porosity (cellulose channels free of lignin) to be easy filled up by a suitable polymer, as well as a suppressor of light absorption. Thus, immersion OC of wood slices by using epoxy resin as an OCA and H_2O_2 steam delignification as an enhancer of OCA permeability is a new general synthetic method for designing of transparent wood composites [28].

Immersion OC can be used as a method for the measurements of RI of tissue components (see Sect. 5.1.4). In application to wood tissue investigation, RI of thermally modified Scots pine wood was obtained by introducing pine wood powder into different OCAs with a stepwise change of RI and measuring light backscattering for a particular wavelength. A number of mixtures of acetone ($n_a = 1.358$) and methylene iodide ($n_b = 1.730$) were used as OCAs with different ratio of their concentrations [29]. The angle of backscattering was 29.4° and the wavelength of the light was 589.6 nm. The RI of untreated Scots pine wood powder was measured as 1.553 and for thermally modified wood at 180 °C, 200 °C, and 230 °C as 1.61, 2.19, and 2.77, respectively.

Remarkably, the transparent wood with embedded molecules of Rhodamine 6G dye allows one to get laser emission due to optical feedback realized within cellulose fibers, which play the role of tiny optical resonators [27].

The tissue OC technique is effective not only for the UV, visible, and NIR ranges but also for much longer waves, such as terahertz (THz) radiation, the transport of which in tissues depends on the content of free and bound water [30–38]. Therefore, the use of hyperosmotic OCAs leads to effective dehydration of tissues, providing a flow of free and weakly bound (mobile) water from the tissue. In turn, the OCA diffuses in the tissue and replaces the mobile water in the interstitial space and cells. As a rule, many OCAs, including glycerol, PG, EG, PEG-200, PEG-300, and PEG-400, have much less absorption than water [36]. Both mechanisms give an increase in the THz wave penetration depth and in the contrast of the THz images [30–38].

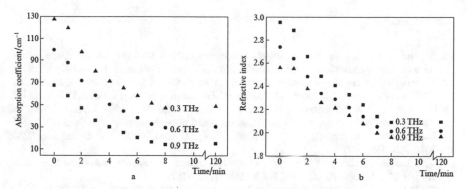

Fig. 9.2 Kinetics of absorption coefficient (**a**) and refractive index (RI) (**b**) of bovine muscle tissue treated with glycerol measured at frequencies of 0.3, 0.6, and 0.9 THz. Time corresponds to the cumulative exposure of the sample to the agent. The nature of the change in the absorption coefficient and RI reflects a decrease in the content of mobile water. (Reprinted with permission from Ref. [32])

Tissue dehydration, i.e., THz clearing, can be also provided by mechanical compression, freezing, heating, gelatin or paraffin embedding, or lyophilization of soft tissue samples [37, 38]. However, most of these methods cannot be used in vivo, as an immersion OC, which can be applied in vivo for reversible tissue dehydration. In Refs. [30–34], glycerol, PG, and PEGs were used as THz immersion OCAs. The decrease of the absorption coefficient and RI of muscle tissue with the time of OCA application was received (see Fig. 9.2a). It is well seen that these common OCAs are efficient providing up to two- to fivefold reduction of absorption of muscle tissue in the frequency range from 0.3 to 0.9 THz during only 8 min of immersion agent topical application. Additionally, a better THz radiation transmittance is caused by a lesser Fresnel reflection from the tissue-air interface as the RI is decreased at OC, for example, from approx. 3.0–2.6 to 2.1–1.9 in the 0.3–0.9 THz range (see Fig. 9.2b).

For healthy and cancerous skin of rats, among anhydrous glycerol, 40% glucose solution, PG, or PEG-600, the latter showed the most effective THz clearing [32–34]. In addition, for cancerous tissue, clearing was faster than for healthy tissue, and for in vivo studies, it was faster (only 13–17 min) compared with in vitro measurements (about 20–80 min). However, for the in vivo case, there is a subsequent process of tissue rehydration due to the physiological response of living tissue to dehydration and osmotic stress.

In Ref. [35], it was shown that a deeper penetration of THz waves into mouse skin, measured in vitro with glycerol treatment, is associated with a decrease in the ratio of free and bound water, studied by MRI.

Thus, the OC technique, which has been approved for use in THz range and can significantly reduce the absorption coefficient and RI of tissues, as well as increase the difference between THz reactions of normal and pathological tissues, opens up new possibilities in Biophotonics [35–38].

References

1. W. Spalteholz, *Über das Durchsichtigmachen von menschlichen und tierichen Präparaten und seine theoretischen Bedingungen, nebst Anhang: Über Knochenfärbung* (S. Hirzel, Leipzig, 1911)
2. W. Spalteholz, *Über das Durchsichtigmachen von menschlichen und tierichen Präparaten und seine theoretischen Bedingungen, nebst Anhang: Über Knochenfärbung* (S. Hirzel, Leipzig, 1914)
3. E.A. Genina, A.N. Bashkatov, Y.P. Sinichkin, I.Y. Yanina, V.V. Tuchin, Optical clearing of tissues: benefits for biology, medical diagnostics, and phototherapy, in *Handbook of Optical Biomedical Diagnostics*, ed. by V. V. Tuchin, vol. 2, 2nd edn., (SPIE Press, Bellingham, WA, 2016), pp. 565–637
4. V.V. Tuchin, *Tissue Optics: Light Scattering Methods and Instruments for Medical Diagnosis*, vol PM 254, 3rd edn. (SPIE Press, Bellingham, WA, 2015)
5. T. Vo-Dihn (ed.), *Biomedical Photonics Handbook* (CRC Press, Boca Raton, FL, 2014)
6. L.V. Wang, H.-I. Wu, *Biomedical Optics: Principles and Imaging* (Wiley-Interscience, Hoboken, NJ, 2007)
7. R. K. Wang, V. V. Tuchin (eds.), *Advanced Biophotonics: Tissue Optical Sectioning* (CRC Press, Taylor & Francis Group, London, 2013)
8. V.V. Tuchin, Optical immersion as a new tool for controlling the optical properties of tissues and blood. Laser Phys. **15**, 1109–1136 (2005)
9. V.V. Tuchin, *Optical Clearing of Tissues and Blood* (SPIE Press, Bellingham, WA, 2006)
10. K. Tainaka, T.C. Murakami, E.A. Susaki, C. Shimizu, R. Saito, K. Takahashi, A. Hayashi-Takagi, H. Sekiya, Y. Arima, S. Nojima, M. Ikemura, T. Ushiku, Y. Shimizu, M. Murakami, K.F. Tanaka, M. Iino, H. Kasai, T. Sasaoka, K. Kobayashi, K. Miyazono, E. Morii, T. Isa, M. Fukayama, A. Kakita, H.R. Ueda, Chemical landscape for tissue clearing based on hydrophilic reagents. Cell Rep. **24**, 2196–2210 (2018)
11. A.A. Selifonov, V.V. Tuchin, Kinetics of optical properties on selected laser lines of human periodontal gingiva when exposed to glycerol-propylene glycol mixture, in International Symposium FLAMN-19 (Fundamentals of Laser Assisted Micro- & Nanotechnologies), Symposium Program, Paper PS3-C02-9, St. Petersburg, 30 June–4 July, 2019, p.71. https://flamn.ifmo.ru/docs/Program_Symposium_FLAMN_-_19.pdf
12. https://www.rxlist.com/lactulose-solution-drug.htm. Accessed 22 Jul 2019
13. K. Iijima, T. Oshima, R. Kawakami, T. Nemoto, Optical clearing of living brains with MAGICAL to extend in vivo imaging. bioRxiv, 507426 (2019). https://doi.org/10.1101/507426
14. S.P. van Rosendal, M.A. Osborne, R.G. Fassett, J.S. Coombes, Guidelines for glycerol use in hyperhydration and rehydration associated with exercise. Sports Med. **40**(2), 113–129 (2010)
15. N.J. Yang, M.J. Hinner, Getting across the cell membrane: an overview for small molecules, peptides and proteins. Methods Mol. Biol. **1266**, 29–53 (2015)
16. D.K. Tuchina, R. Shi, A.N. Bashkatov, E.A. Genina, D. Zhu, V.V. Tuchin, Ex vivo optical measurements of glucose diffusion kinetics in native and diabetic mouse skin. J. Biophotonics **8**(4), 332–346 (2015)
17. D.K. Tuchina, A.N. Bashkatov, A.B. Bucharskaya, E.A. Genina, V.V. Tuchin, Study of glycerol diffusion in skin and myocardium ex vivo under the conditions of developing alloxan-induced diabetes. J. Biomed. Photon. Eng **3**(2), 020302 (2017)
18. J. Wang, Y. Zhang, T.H. Xu, Q.M. Luo, D. Zhu, An innovative transparent cranial window based on skull optical clearing. Laser Phys. Lett. **9**(6), 469–473 (2012)
19. E.A. Genina, A.N. Bashkatov, O.V. Semyachkina-Glushkovskaya, V.V. Tuchin, Optical clearing of cranial bone by multicomponent immersion solutions and cerebral venous blood flow visualization. Izv. Saratov Univ. (N. S.), Ser. Phys. **17**, 98–110 (2017)
20. Y.-J. Zhao, T.-T. Yu, C. Zhang, Z. Li, Q. Luo, T.-H. Xu, D. Zhu, Skull optical clearing window for in vivo imaging of the mouse cortex at synaptic resolution. Light Sci. Appl. **7**, e17153 (2018)

21. E.A. Susaki, H.R. Ueda, Whole-body and whole-organ clearing and imaging techniques with single-cell resolution: toward organism-level systems biology in mammals. Cell Chem. Biol. **23** (1), 137–157 (2016)
22. W. Zhang, S. Liu, W. Zhang, W. Hu, M. Jiang, A. Tamadon, Y. Feng, Skeletal muscle CLARITY: a preliminary study of imaging the three-dimensional architecture of blood vessels and neurons. Cell J. **20**(2), 132–137 (2018)
23. E. Lee, J. Choi, Y. Jo, J.Y. Kim, Y.J. Jang, H.M. Lee, S.Y. Kim, H.-J. Lee, K. Cho, N. Jung, E.M. Hur, S.J. Jeong, C. Moon, Y. Choe, I.J. Rhyu, H. Kim, W. Sun, ACT-Presto: rapid and consistent tissue clearing and labeling method for 3-dimensional (3D) imaging. Sci. Rep. **6**, 18631 (2016)
24. C.A. Warner, M.L. Biedrzycki, S.S. Jacobs, R.J. Wisser, J.L. Caplan, D.J. Sherrier, An optical clearing technique for plant tissues allowing deep imaging and compatible with fluorescence microscopy. Plant Physiol. **166**, 1684–1687 (2014)
25. K.J.I. Lee, G.M. Calder, C.R. Hindle, J.L. Newman, S.N. Robinson, J.J.H.Y. Avondo, E.S. Coen, Macro optical projection tomography for large scale 3D imaging of plant structures and gene activity. J. Exp. Bot. **68**(3), 527–538 (2017)
26. Visikol, Inc, *White Paper: Clearing Agent for Botanical Microscopy* (Visikol, Inc, Whitehouse Station, NJ, 2016). Accessed at www.visikol.com
27. E. Vasileva, Y. Li, I. Sychugov, M. Mensi, L. Berglund, S. Popov, Lasing from organic dye molecules embedded in transparent wood. Adv. Optical Mater **5**, 1700057 (2017)
28. H. Li, X. Guo, Y. He, R. Zheng, A green steam-modified delignification method to prepare low-lignin delignified wood for thick, large highly transparent wood composites. J. Mater. Res. **34**(6), 932–940 (2019)
29. I. Niskanen, J. Heikkinen, J. Mikkonen, A. Harju, H. Herajarvi, M. Venalainen, K.-E. Peiponen, Detection of the effective refractive index of thermally modified Scots pine by immersion liquid method. J. Wood Sci. **58**, 46–50 (2012)
30. M.M. Nazarov, A.P. Shkurinov, E.A. Kuleshov, V.V. Tuchin, Terahertz time-domain spectroscopy of biological tissues. Quant. Electron. **38**(7), 647–654 (2008)
31. M. Nazarov, A. Shkurinov, V.V. Tuchin, X.-C. Zhang, Terahertz tissue spectroscopy and imaging, in *Handbook of Photonics for Biomedical Science*, ed. by V. V. Tuchin, (CRC Press, Taylor & Francis Group, London, 2010), pp. 591–613. Chapter 23
32. A.S. Kolesnikov, E.A. Kolesnikova, A.P. Popov, M.M. Nazarov, A.P. Shkurinov, V.V. Tuchin, In vitro terahertz monitoring of muscle tissue dehydration under the action of hyperosmotic agents. Quant. Electron. **44**(7), 633–640 (2014)
33. A.S. Kolesnikov, E.A. Kolesnikova, D.K. Tuchina, A.G. Terentyuk, M. Nazarov, A.A. Skaptsov, A.P. Shkurinov, V.V. Tuchin, In-vitro terahertz spectroscopy of rat skin under the action of dehydrating agents. Proc. SPIE **9031**, 90310D (2014). https://doi.org/10.1117/12.2052226
34. A.S. Kolesnikov, E.A. Kolesnikova, K.N. Kolesnikova, D.K. Tuchina, A.P. Popov, A.A. Skaptsov, M.M. Nazarov, A.P. Shkurinov, A.G. Terentyuk, V.V. Tuchin, THz monitoring of the dehydration of biological tissues affected by hyperosmotic agents. Phys. Wave Phenom. **22**(3), 169–176 (2014)
35. O. Smolyanskaya, I. Schelkanova, M. Kulya, E. Odlyanitskiy, I. Goryachev, A. Tcypkin, Y.V. Grachev, Y.G. Toropova, V. Tuchin, Glycerol dehydration of native and diabetic animal tissues studied by THz-TDS and NMR methods. Biomed. Opt. Express **9**(3), 1198–1215 (2018)
36. G.R. Musina, A.A. Gavdush, D.K. Tuchina, I.N. Dolganova, G.A. Komandin, S.V. Chuchupal, O.A. Smolyanskaya, O.P. Cherkasova, K.I. Zaytsev, V.V. Tuchin, A comparison of terahertz optical constants and diffusion coefficients of tissue immersion optical clearing agents. Proc. SPIE **11065**, 468 (2019). https://doi.org/10.1117/12.2526168. Saratov Fall Meeting 2018: Optical and Nano-Technologies for Biology and Medicine

37. O.A. Smolyanskaya, N.V. Chernomyrdin, A.A. Konovko, K.I. Zaytsev, I.A. Ozheredov, O.P. Cherkasova, M.M. Nazarov, J.-P. Guillet, S.A. Kozlov, Y.V. Kistenev, J.-L. Coutaz, P. Mounaix, V.L. Vaks, J.-H. Son, H. Cheon, V.P. Wallace, Y. Feldman, I. Popov, A.N. Yaroslavsky, A.P. Shkurinov, V.V. Tuchin, Terahertz biophotonics as a tool for studies of dielectric and spectral properties of biological tissues and liquids. Prog. Quant. Electron. **62**, 1–77 (2018)
38. K.I. Zaytsev, I.N. Dolganova, N.V. Chernomyrdin, G.M. Katyba, A.A. Gavdush, O.P. Cherkasova, G.A. Komandin, M.A. Shchedrina, A.N. Khodan, D.S. Ponomarev, I.V. Reshetov, V.E. Karasik, M.A. Skorobogatiy, V.N. Kurlov, V.V. Tuchin, Malignancy diagnosis using terahertz spectroscopy and imaging: a review. J. Opt. Accepted Manuscript online 15 October 2019

Index

© The Author(s), under exclusive license to Springer Nature Switzerland AG 2019
L. M. C. Oliveira, V. V. Tuchin, *The Optical Clearing Method*,
SpringerBriefs in Physics, https://doi.org/10.1007/978-3-030-33055-2

Printed in the United States
By Bookmasters